Vision, Illusion and Perception

Volume 4

The Vision, Illusion and Perception (VIP) book series publishes new developments and advances in the fields of Vision and Perception research, rapidly and informally and with a high quality. The series publishes fundamental principles as well as state-of-the-art theories, methods and applications in the highly interdisciplinary field of Vision Science, Perception and multisensory processes related to vision. It covers all the technical contents, applications, and multidisciplinary aspects of fields such as Cognitive Science, Computational and Artificial Intelligence, Machine Vision, Psychology, Physics, Eye Research, Ophthalmology, and Neuroscience. In addition, the series will embrace the growing interplay between the art and science of vision. Within the scope of the series are monographs, popular science books, and selected contributions from specialized conferences and workshops.

Jan B. Deręgowski

On Picture Making and Picture Seeing

A Brief Discourse

Springer

Jan B. Deręgowski
School of Psychology
University of Aberdeen
Scotland, UK

ISSN 2365-7472 ISSN 2365-7480 (electronic)
Vision, Illusion and Perception
ISBN 978-3-031-23347-0 ISBN 978-3-031-23348-7 (eBook)
https://doi.org/10.1007/978-3-031-23348-7

This Springer imprint is published by the registered company Springer Nature Switzerland AG
The registered company address is: Gewerbestrasse 11, 6330 Cham, Switzerland

Preface

The genesis of this wee book needs to be made public so that responsibility for its very existence can be justly proportioned.

I was walking up the back staircase to my room, humming a bucolic ditty about grasshoppers, butterflies and springtime, when I noticed Prof. Benjamin W. Tatler lying in wait on a landing above, in a position of undoubted strategic superiority. He spoke: "Would you like to write a book?"

It clearly was a cunning, Machiavellian, diversionary ploy intent to disrupt my peaceful routine activities and banish the, entirely illusory, threat of my invasion of his laboratory space.

Quick as a lightening I counter-attacked: "Will you read it?" (My stratagem was: He is bound to smile and say "Certainly not!", to which I will reply with the pleasantest of smiles: "Then there is no point in writing it.")

"Yes", he answered.

My counterattack had failed. Here is the consequence.

Before the present condition could be reached several other agents became, of necessity, involved in the book's lame progress. Chief of them was Mrs Eva Loft Deręgowska, who without carping, grousing, moaning or whining, typed and repeatedly retyped the text. Mr Kenneth Le Riche, an artist and lecturer in art, patiently and graciously discussed with me various issues touching on the problems of artistic creativity and gently pointed out gaping lacunae in my knowledge; and after reading a draft of one of the chapters, made a most laudatory and gratifying comment any author has ever heard: "I have never seen anything like it!" As a mark of gratitude for these honeyed words, handprints of both his daughters have been incorporated (see Fig. 1.2B).

Dr. Helen Saunderson, a psychologist-cum-artist and a colleague, as too polite to dismiss my plea that she should read the text and Mr Jim Urquhart, the *eminence grise* of matters technical in the spinney that I inhabit, did his bit. All of them must share the responsibility.

When the text was complete, nothing followed. There were no flourishes of trumpets, no bells tolled. There was silence. Pestilence, called Covid 19, had descended on the world and the University was closed. It occurred to me that another opinion on the text should be sought, so I wrote to Prof. Parker, my sometime colleague who was now spying on warblers in the wetland of Norfolk. His reply told me that Savi's warbler, a rarity, had been seen, yet in spite of this he would read the text and let me have his comments. When they came, I treated them with great respect, dismissing only those that were contrary to my notion. Prof. Parker is therefore yet another person responsible for the book's existence.

Here is the book. It is about the making and seeing of pictures. It is a book about perception. Perception furnishes skeletons for artistic creations.

A text presuming to treat on pictorial matters inevitably calls for many pictures. There's the rub. Fortunately, Mr Peter Bates agreed to come out of his retirement and to prepare all the line drawings and, here is surprise, Prof. B. Tatler (driven by qualms of conscience, who knows) offered to take overall care of figures. May they both age gracefully.

Many of the figures are reproduced with gracious permission by the copyright holders. Their generosity is acknowledged in the figure legends.

Scotland, UK Jan B. Deręgowski

Contents

About the Author

Jan B. Deręgowski D.Sc., FRSE, FBPsS, is Professor Emeritus of Psychology at the University of Aberdeen. He has been pursuing studies of visual perception in many cultures and has published some of his findings in *Illusions, Patterns and Pictures* (Academic Press, 1980) as well as, recently, two overviews: *Cross-cultural Studies of Illusions* (in the Oxford compendium of Visual Illusions, OUP, 2017) and *The Psychology of Graphic Perception* (in the Oxford Handbook of Archaeology and Anthropology of Rock Art, OUP, 2018).

Chapter 1
Introduction in Which the Reader Enters Pictorial Donnybrook

When Zeuxis Primus was suddenly inspired to draw a bison he saw the world differently. Previously it was filled with bisons, men, trees—coherent entities whose minutiae were scarcely noticeable. In such a world Zeuxis Primus used his vision to the same purpose as do all creatures: to make sense of the environment, to identify objects and their locations in space. Indeed, so great is vision's keenness to achieve this end that sometimes it yields to guile. Men make decoys to entrap animals and other men. Other creatures are by nature endowed with the wherewithal of dissimulation. The matamata turtle of Brazilian rivers has about its mouth and throat a series of fleshy red filaments, some of which are voluntarily movable whilst more slender ones sway with every movement of water. Their colour and worm-like form and motion serve as lures to attract fishes and frogs.[1] Matamata thus does what anglers do to catch fish and as street artists do to catch the eye of a potential purchaser. Zeuxis would like his picture to beguile men as convincingly as the appendages of the Matamata's gob beguile unfortunate fishes and frogs.

A successful depiction, in Zeuxis's view, is *one* which causes *a viewer* to err, to perceive something which does not really exist, but has sufficient visual characteristics to enable a viewer to identify it as something real. It shows the most visually salient features of objects and mitigates the impact of those features which the model shares with other potential models. Zeuxis's tools are robust but simple, *and* of these the stylus, with which he can scratch lines on the rock face, is the most important.

In order to succeed Zeuxis Primus has to detect telling features which can be represented by scratched lines. Most of the lines which he sees are figments of his visual system. They do not exist in the way that lines on a ruled copy book page exist. They are ephemeral lines running along the boundaries of surfaces or parts of visual entities, such lines are subject to change when the viewpoint changes or the object changes or moves. One such particularly salient line is that along the boundary defining the extent of an object—the object's peripheral contour—a temporary outline seen against a background e.g. the outline of a cow seen against the background of the meadow it grazes in. This outline is subject to change, not only with the object viewed, but also, generally but not inevitably, with the object's

J. B. Deręgowski, *On Picture Making and Picture Seeing*,
Vision, Illusion and Perception 4, https://doi.org/10.1007/978-3-031-23348-7_1

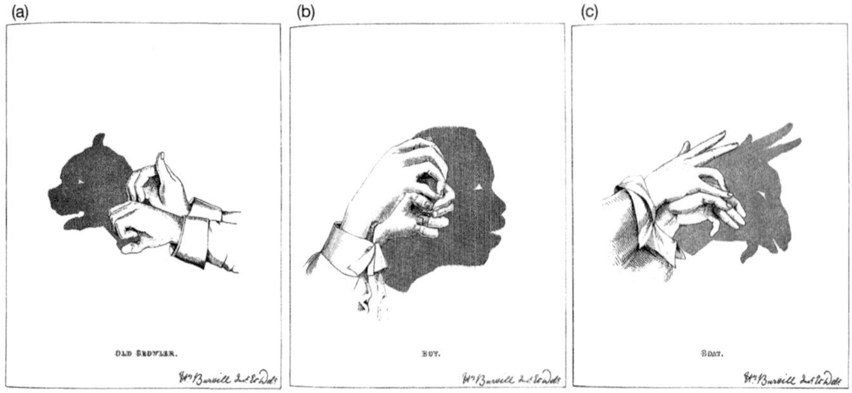

Fig. 1.1 Three of the shadow figures which can be cast by human hands as ingeniously shown by Mr. Henry Bursill (1859): **a** Growler the dog, **b** a boy, **c** a goat; all defined by their peripheral contours

orientation; a ball's peripheral contour is always circular, that of a postcard changes with the slightest change of orientation.

When the peripheral contour is disrupted or obliterated and its visibility fades, so does the visibility of the object, a fact exploited by users of camouflage ranging from zebras and copperhead snakes to the wearers of military battle uniforms. The presence of a peripheral contour does indicate the presence of an entity but does not necessarily identify it—human hands can project the peripheral contours of a variety of objects, of, for example, a dog, a boy, and a goat (Fig. 1.1), as Mr. Bursill[2] has ingeniously shown. Hands can also project a strikingly veridical peripheral contour, as shown by handprints, designs which have decorated ancient caves and are still joyfully made in today's nurseries (Fig. 1.2). When hands are used as stencils, held close to the surface and dye is blown at them, the stencilled shape can be varied by bending fingers. Pictures of distorted hands can thus be made. Such Palaeolithic pictures are sometimes found and sometimes said to be of amputee's hands. This seems to be a misattribution. It seems more likely that they are relics of a Palaeolithic parlour game.[3]

Both the shadow cast in accordance with Mr. Bursill's instructions and the handprints are optically legitimate projections of hands. A knowing observer is surprised by the Bursillian shadows because they are unexpected, but a viewer ignorant of the manner in which the figures were made might accept all of them for what they appear to be. The perceptual system is eager to make sense of such minimal cues as those of peripheral contours, even of peripheral contours of random ink blots.[4]

The ambiguity of perceptual contours exemplified by Mr. Bursill's figures can be moderated by further cues *within* the peripheral compass.

These cues, *the significant contours*, just like the peripheral contours, are provided by the boundaries between visually distinct areas, but they do not encompass objects. These encompassed lines may be created by changes of hue in the surface, reflectivity,

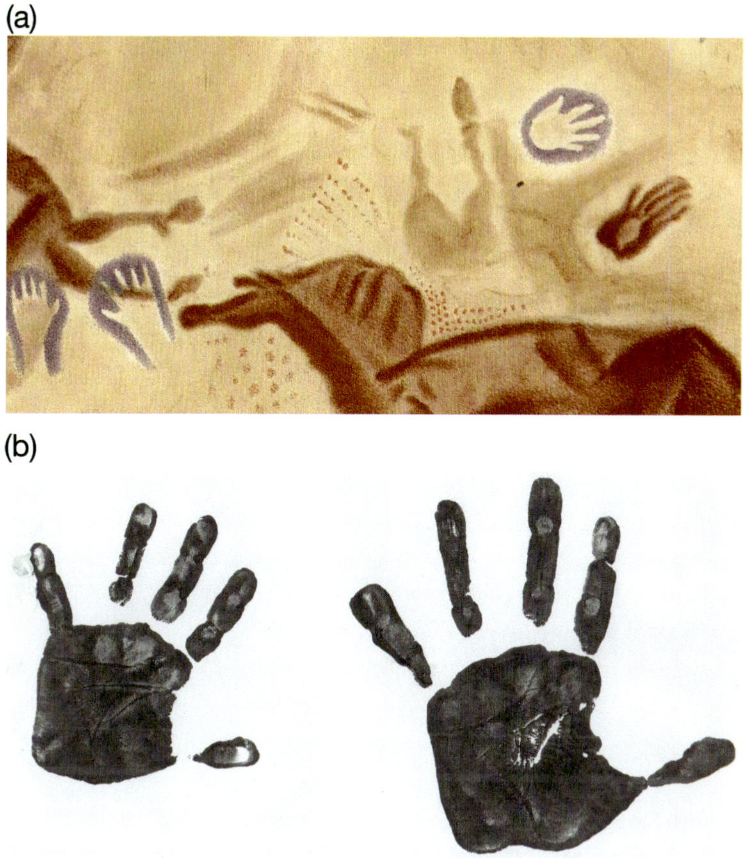

Fig. 1.2 Two kinds of hand prints have been used since times immemorial. Positive handprints are made by pressing a palm covered with a pigment against a surface; spraying of dye on a hand pressed to the surface yields a negative handprint. Both kinds were found, side by side, by Abbé Breuil in the Altamira Cave (Breuil and Obermeier 1935) (**a**); modern positive handprints (**b**) are by Malina and by Halszka Le Riche

or texture gradient brought about either by the change of its shape or by some external agent such as light.

Significant contours make identification of the percept more likely; more likely but not certain, since, as the encounters of frogs and fishes with matamata turtle show, one can never be certain.

Figure 1.3 illustrates this point. It is called Mach's Book after the Moravian scientist, he of the Mach's Number, Mach's Angle, and Mach's Principle and, which may be of particular interest to artists, Mach Bands.[5] When (a) only its peripheral contour is shown it appears flat, when (b) a significant contour is introduced it appears three-dimensional, but its three-dimensionality is of an uncertain kind for the spine of the book appears now near, now far from the viewer.

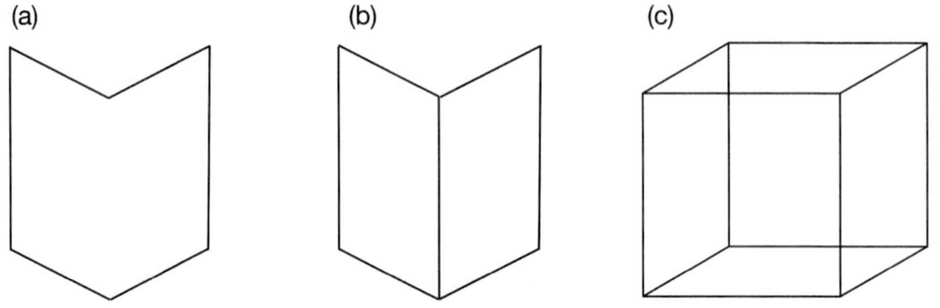

Fig. 1.3 Mach's book (1897); **a** peripheral contour of the book. In **b** the peripheral contour is augmented by the contour of the spine; **c** the Necker Cube (Necker, 1832)

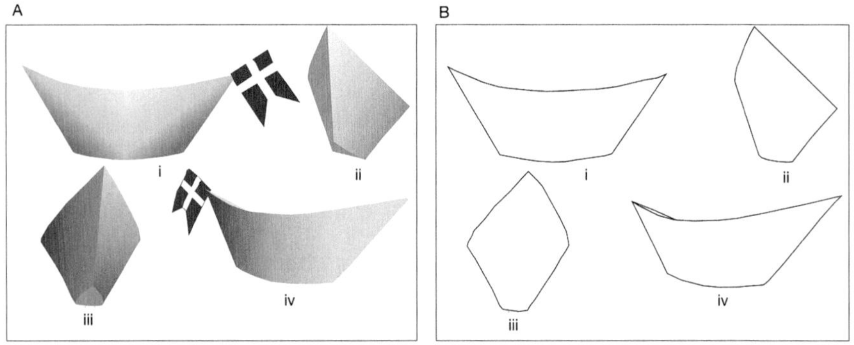

Fig. 1.4 **A** Four aspects of a small paper boat, showing changes of the peripheral contour as the boat turns around: (i) Peripheral contour is made of the typical contour and shows the bow and the stern. (ii) Peripheral contour is not identical with typical contour; part of the typical contour (the bow) forms a significant contour. (iii) The significant contour of the bow forms axis of symmetry. (iv) As in (i) but the boat is facing other way. **B** Peripheral contours of the boat in the four orientations

The reader may well ask why *this* book shows no such fluctuations. It does not because its pages present visual cues not found in the Mach's book. The edges of its open pages and the lines of print parallel to them, all seem to converge with distance from the eye. The surface of the page subtly changes its appearance. These gradients strongly hint that the significant contour, which the two inclined pages form, runs in a valley.

A wee paper boat in Fig. 1.4A elucidates the complex relationship among contours. In figure (i) a side view of the boat is shown incorporating views of both its bow and stern (which happen to be identical). As the boat turns its stern disappears and the contour role of the bow changes from that of periphery to significance (figure ii). On further rotation the bow passes through the orientation in which it forms the axis of symmetry of the display (figure iii) only to resume its function as a part of the peripheral contour (figure iv). Peripheral contours which these views present are summarised in Fig. 1.4B.

The ranges of angles within which particular shapes of peripheral contours are seen as permanent characteristics of the boat are variable. Within each of the ranges, what psychologists call shape constancy prevails and ensures that "whatever the angle from which we observe an object, it preserves its solid and permanent shape".[6] It does so in spite of changes of the retinal image: shape constancy is a higher order aspect of visual analysis that maintains the integrity of the object despite the moment to moment changes caused by rotation of the object or by a changing viewpoint. Two such ranges dominate perception of a turning boat: one, the wider of the two is centred on the boast's side view (Fig. 1.4(i) and (iv)) and the other, much narrower, on its bow/stern views (Fig. 1.4(iii)). As the boat is doubly symmetrical its two side views are identical, as are the views of the bow and the stern. This does not apply to a bison (the subject to be addressed shortly), whose front view, as every bison knows, differs from rear view and who does therefore, on rotation, present four different contours, two of which are enantiomorphic. As the complexity of the object increases, so does the number of distinct angular ranges presenting its shape.

Peripheral contours of an object may differ greatly in the extent to which they look like the object. Only two of the contours depicted are (in Fig. 1.4B), prima facie, boat-like. A change of the peripheral contour may change appearance radically. The fact clearly, but unknowingly, one suspects, acknowledged by a panicking pangolin, who on rolling itself up not merely creates a barrier of impenetrable horny plates, but also changes the peripheral contour from that of a pangolin to that of a spheroid; a most unpangolinian, circular, peripheral contour, which remains circular whatever the stance of the observer. Its visual identity vanishes. The three-banded Brazilian armadillo uses a similar emergency procedure, as does the humble and unexotic hedgehog. The peripheral contour of the hedgehog may be more or less hedgehoggy. When it is at its hedgehoggist it is the *typical* contour—the blazon of all hedgehogs.

Comparison of the depicted contours shows that there are ranges of contours readily recognisable as derived from the same object. Peripheral contours, referred to here as *typical* contour; the emblem of all hedgehogs are observable only when a specific spatial relationship between the observer and the object prevails, as when the object is viewed from a specific stance or set at a specific angle. In the case of the wee boat a range of typical contours appears to centre on the first view shown in Fig. 1.4A; in many animals it does so about the plane containing their spine and is seen most clearly when it lies in the observer's fronto-parallel plane. The peripheral contours contained within these ranges are the *typical contours* and proffer the most lucid representations of the objects from which they are derived, as attested by the animals painted on the ceiling of the Altamira Cave (Fig. 1.5).[7] Such contours of different animals portrayed on the same wall are by their very nature co-planar. The resulting picture of the arrangement lacks pictorial depth. Attempts to introduce depth are difficult and may lead to loss of consistency.

Such an instance of spatial inconsistency is presented in a French fifteenth century illustration of Boccaccio's De Claris Mulieribus. It shows (Fig. 1.6) young women gathering cocoons and spinning silk threads. The spinner is so engrossed in her task that she is unaware of the impossibility of the frame on which she is working.[8]

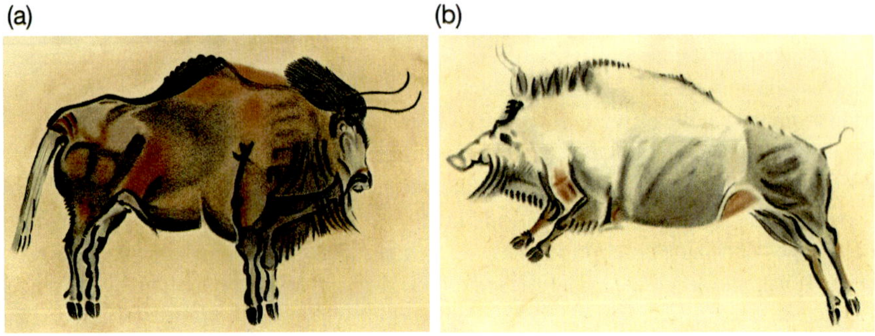

Fig. 1.5 Two animals of the Altamira Cave as copied by Abbé Breuil[7]: a "Stationary Bison" and a "Galloping Wild Boar" (From Breuil and Obermeier (1935), courtesy of University of Aberdeen Library)

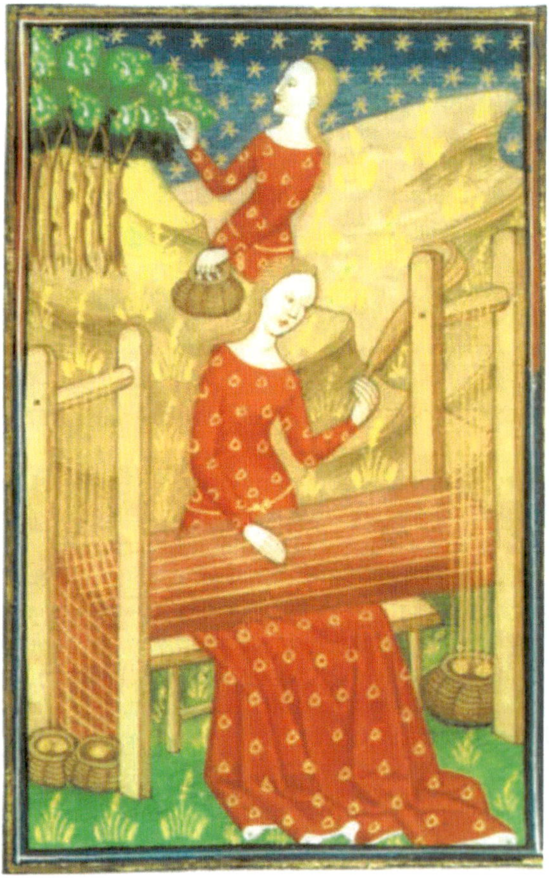

Fig. 1.6 Illustrious ladies spinning silken threads on an awry frame. Illustration of G. Boccaccio's De Claris Mulieribus (15th cent. French). The artist who attends to one feature at a time, with such intensity that he neglects its relationships with other features, may accidentally create as picture in which visually acceptable features combine to form a visually unacceptable entity

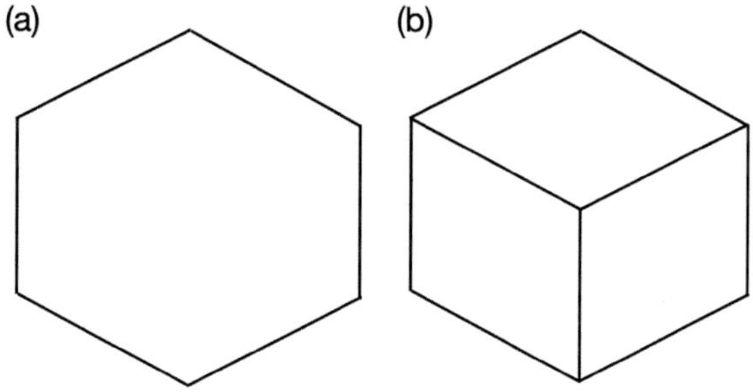

Fig. 1.7 The impact of significant contours, or how a flat hexagon, becomes a cube

Looking at an isolated object in absence of external depth cues, the eye sees the object in the plane perpendicular to the line of sight, the fronto-parallel plane, in which the peripheral contour lies. The object can, however, be imbued with three-dimensionality by depth cues placed within its compass. Thus a circular peripheral contour may, depending on the texture gradient, depict a bird's eye view of a hemispherical bowl or a ball. (We shall return to this phenomenon in the third chapter.)

A flat hexagon can be converted into a cube by insertion of a Y-shape (Fig. 1.7), the acquired significant contours dictating three-dimensionality of the percept.[9] This effect would also be observed if the physical Y-lines were not there, but were implied by the differences among the surface patterns of the three faces of the depicted cube. Such patterns, although of fundamental visual importance, have no generic name in the language of the market place, unlike e.g. outline or shadow. They are the texture gradients and are seen whenever a surface is seen. A pronounced texture gradient is seen, for example, by a man looking along a pebbly foreshore, the more distant pebbles appearing smaller and more closely packed. The rate at which these visual characteristics appear to change with distance is determined by the angle between the line of sight and the ground level, the plane looked at. These changes of appearance translated onto the artist's canvas create the impression of depth. The same perceptual processes are involved in perception of texture gradients as are involved in perception of linear perspective.[10]

When a texture gradient seems unchanging the surface is perceived as in the viewer's fronto-parallel plane, and all the information about it is contained in the peripheral contour—as it is in Mr. Bursill's shadow figures.[2] Such perceptual uniformity of texture gradient is not necessarily associated with physical uniformity, it may also be a result of amalgamation of varied texture gradients within the same peripheral contour. Then the peripheral contour determines the percept i.e. our visual understanding of the object, and when it happens to be the typical contour it dominates.

A perceptual amalgamation of features also takes place when peripheral contours are nearly parallel and close together and thus imply *central lines*. This is so, for example, in the case of flamingos' elegant legs (Fig. 1.8). The Altamira bison's horns are thus depicted as shown in Fig. 1.5. The technique can be readily extended to embrace bulkier models and was used by artists from cultures lacking a tradition

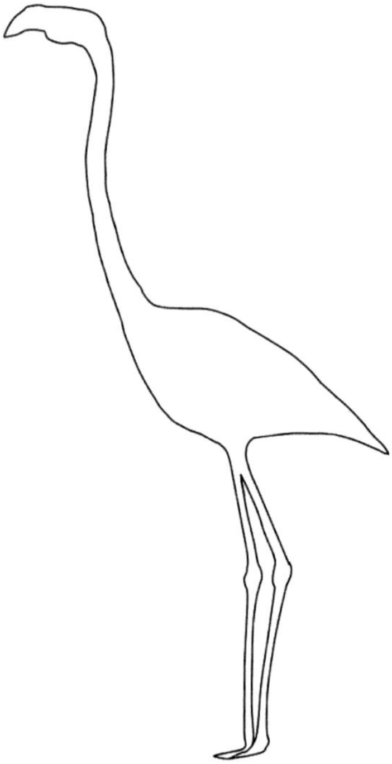

Fig. 1.8 Central lines suffice to portray a flamingo's elegant legs

of picture making, as well as by artists from cultures in which picture making thrived. Drawings by Tallensi,[11] who lacked such a tradition, of a man, a woman, a horse rider and a cow, are done using this technique, and show its limitations. It cannot cope with entities devoid of distinct central lines, such as heads, navels, and pendant breasts (Fig. 1.9a). The inherent difficulty of the task is probably responsible for the scarcity of such delineations in children's drawings.[12] Pictures by South African Bushmen, members of a culture with a rich artistic tradition, show the same technique. Its limitations are boldly exposed by the lady with a raised digging stick, who is broad where the broad should be (Fig. 1.9b).

A related phenomenon is that of inconsistency of perceptual definitions. It is demonstrated here by the two-pronged trident in which contours at the two ends of the figure are incompatible (see Fig. 1.10). The contours on the left form two square prongs, each defined by two peripheral contours and one significant contour. The contours on the right form three round prongs, each defined by two peripheral contours. Two of the contours have clearly perceptually mutated, and the perceptual system is unwilling to accept this mutation (see also the spinning frame, Fig. 1.6).

Analogous mutation can be seen in the rock painting of giraffes and other animals at Aïr in Southern Sahara.[13] The legs of the giraffe shown in Fig. 1.11 can either be seen as defined by peripheral contours (the near legs obscuring those further), or as depicted by means of central lines (four legs showing). It is not possible to tell

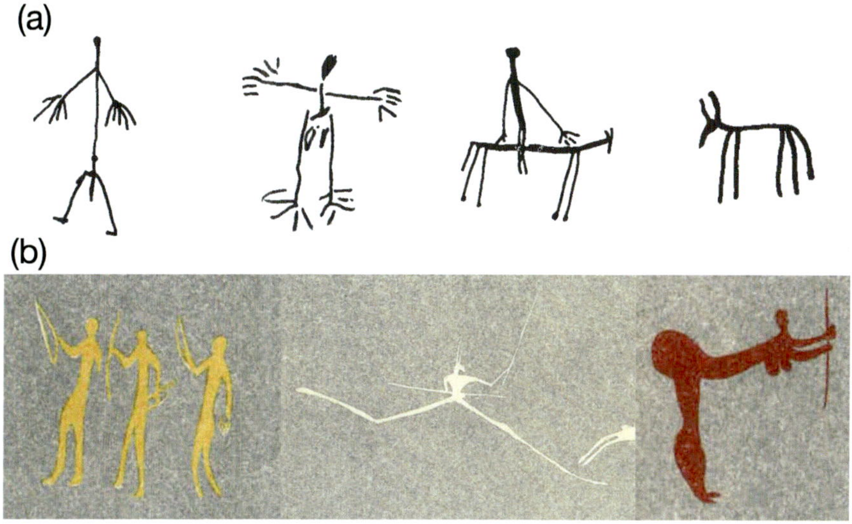

Fig. 1.9 Use of central lines in art: **a** four portrayals by entirely unexperienced Tale draughtsmen, are easily seen as those of a man, a woman, a man on a horse, and a cow. The woman's pendant breasts, which are carried rather low, and the heads of all people show that the artists had to resort to peripheral contours when central lines could not be seen (Fortes 1940). See also Fortes (1981) for a broader discussion. **b** Bushman figures copied from rock shelters by Tongue (1909). The lady with a digging stick testifies that not all parts of all models can be represented by central lines. The superiority of the Bushmen figures over those of the unexperienced Tale draughtsmen is clearly visible

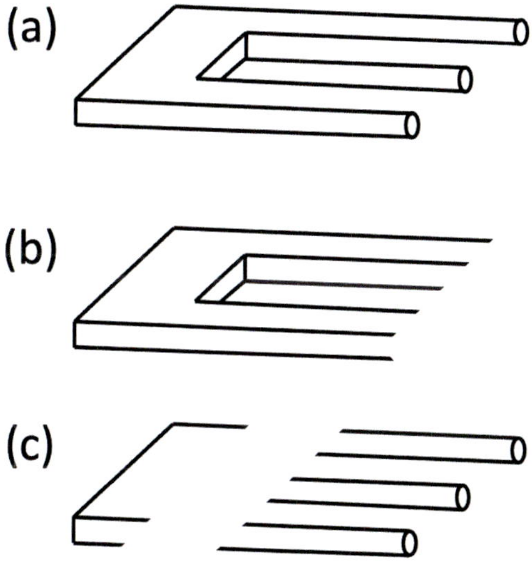

Fig. 1.10 Two-pronged trident, a figure in great vogue in the introductory texts on perception. Figure **a** shows the entire two-pronged trident. Figure **b** shows its two square prongs, each defined by two peripheral contours and a significant contour. Figure **c** shows its three round prongs, each defined by two peripheral contours. The change of the role and grouping of the peripheral (p) and the significant (s) contours from figure (**b**) to figure (**c**) can be described thus: (psp) (psp) into (pp) (pp) (pp)

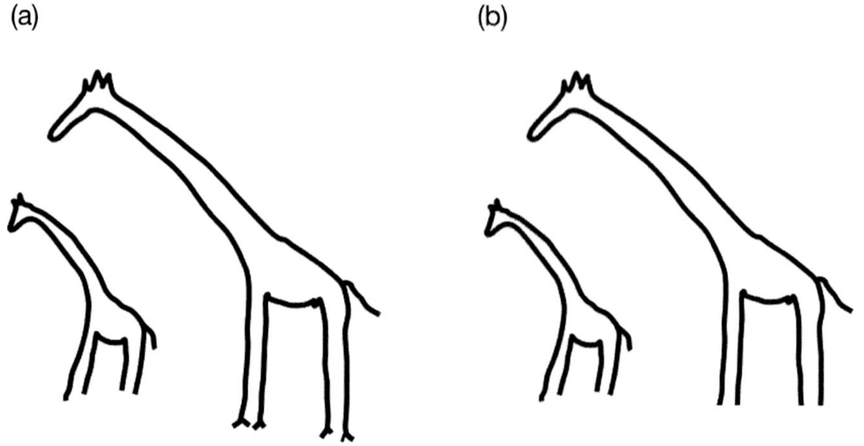

Fig. 1.11 The legs of the giraffes of Aïr confuse most viewers, who know that giraffes have legs and that legs should therefore be depicted, but a picture may show either four legs or, when showing a side view, because of overlap, only two. Looking at the giraffes in the Aïr picture the viewer has to decide. Does it portray four legs by means of central lines, or only the pair of legs closer to the viewer? Figure **a** shows two giraffes as they appear on the rock face: the small "biped" and the large "quadruped" portrayal. In figure **b** the hooves of the larger giraffe have been obliterated and the portrayal is now "biped"?[13] (Based on drawings in Rodd (1938))

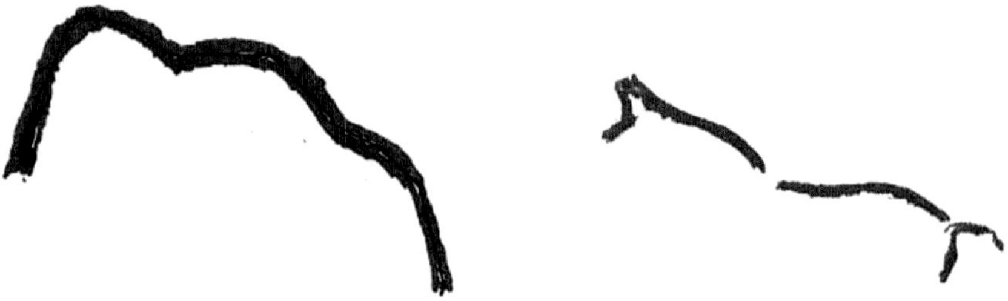

Fig. 1.12 Segments of typical contours can evoke percepts. Their perception, however, is greatly affected by their orientation, as turning the page upside-down will show

whether the artists of Aïr intended this visual game for themselves, for archaeologists, or whether they had been distracted while drawing.

Alone in a cave in the unsteady light of greasy cressets, Zeuxis I is bent on depicting a bison. He has seen many bisons. Now sitting on a stone he recalls their shapes, the texture of their hides, their smells. He cannot depict the smells and the textures but he can depict shapes. He remembers them particularly well, most of all he remembers the typical contours. He gets up and scratches a single line on the rock face, bends down and picks up a cresset to see better what he has drawn. What he has drawn is not a full typical contour. It is only a segment but by this segment the entire animal is brought to mind (Fig. 1.12).

Generally, bodies have more than one typical contour and these contours vary in importance with the purpose of the drawing. The dominant typical contour of the human foot is often ignored when portraying people walking or standing; instead optically less cogent but more befitting the depicted arrangement, a side view, involving less striking contours is presented. The dominant contour runs horizontally along the edge of a foot resting on the ground. It is the contour of the footprint. Its salience is acknowledged universally. It is recognised by the Bali theatre puppet makers,[14] as well as by the Nasca embroidery workers of Peru.[15] The saints of the seventeenth century Ethiopian manuscript stand, rather uncomfortably, on the edges of their feet, as do the Chinese men on the carved boards of Gosham[16] (Fig. 1.13). The range of diverse cultures in which such portrayals are found argues that ***the*** typical contour is noticed but that it is disregarded in many pictures in favour of the lateral view of the feet because it lies in a plane orthogonal to that of the upright leg and the artist is concerned with the model as a whole, and so the leg is more important than the foot.

There are pictures, particularly rock pictures, in which typical contours of hooves and paws of animals appear under their bodies. These pictures show typical contours of hooves and paws, pendant from the legs. The Tin Wana elephant is a weighty example of this kind of usage. Its legs are portrayed by central lines, but at the end of each central line dangles a circle, a semblance of a pad (Fig. 1.14a). This is also so, in the case of the jolly, exuberant lion of Tyfelfontein, painted by Bushmen of Namibia[17] and shown here in Fig. 1.14b. It has two pendant paws. It also has a paw at the tip of its tail—testimony to the power of typical contours. This surprising appendage testifies to a sense of humour and contrasts with the notion popular with some humourless students of rock art that typical contours of paws and hooves are drawn to identify species. This manner of representation driven by the typicality of contour is also responsible for depiction of rectangular garden pools and rugs laid on the ground by 'pendant' rectangles in oriental art (Fig. 1.15).

Disparity between the orientations of the plane of the typical contour of the foot and that of the peripheral contours of the leg, a disparity which the lion of Tyfelfontein so ably illustrates, occurs in much more pronounced form in objects consisting of elements having non-coplanar typical contours. Australian crocodiles display this very characteristic and Australian artists are adept at solving the graphical problem which it presents.

A crocodile at rest appears a tripartite animal with distinct head, trunk and tail. The head and the tail have their typical contours in the vertical plane. The typical contour of the trunk lies in the horizontal plane. Placing these contours in the same plane, the plane of the picture, distorts the spatial relationship between them but ensures that each of them is presented in its most telling way. The single notional viewing stance is abandoned but the meaning of elements is preserved (Fig. 1.16).

The following conversation between an Australian artist and an European visitor could well take place:

AA: What do you see here?
EV: A crocodile, but it is wrong.
AA: What is wrong? Is its head wrong?

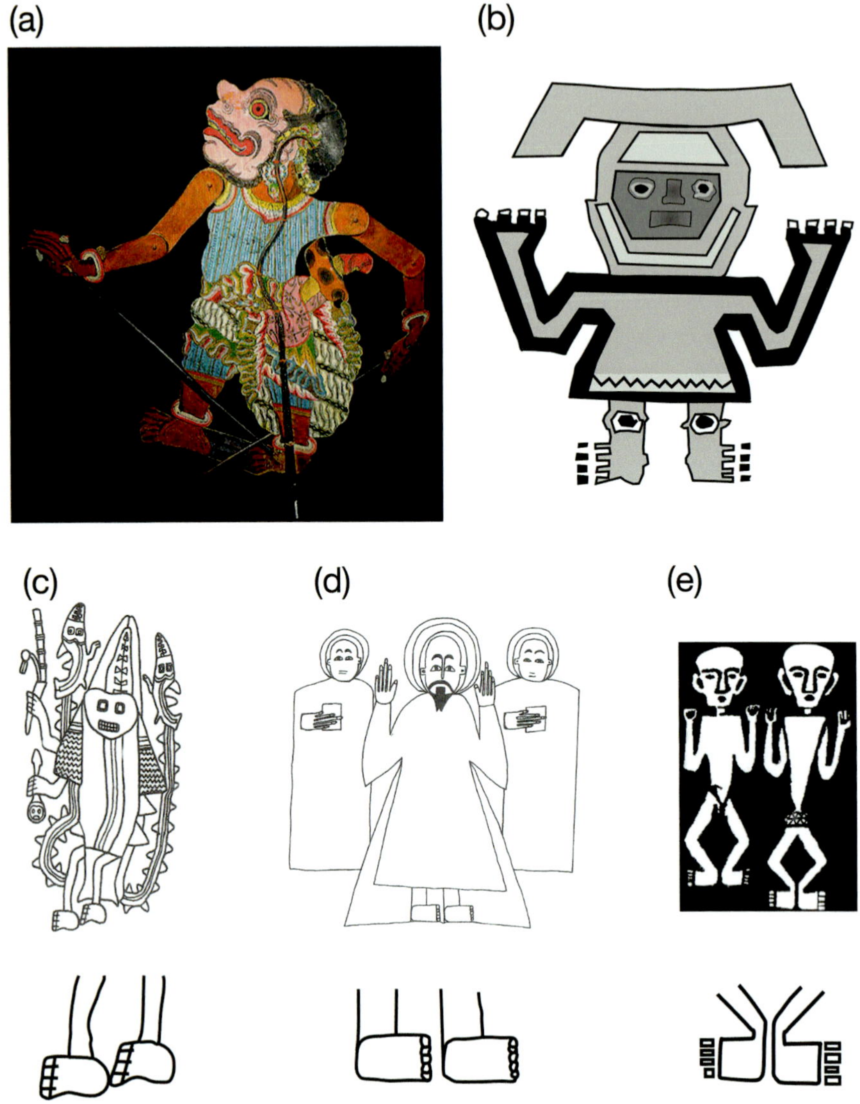

Fig. 1.13 Feet throughout space and time: **a** Bali, a puppet from a shadow theatre, **b** and **c** Peru, ceremonial cloths, **d** Ethiopian (17th cent.) book of Gospels, St. Mark, **e** China, wood engraving (*Sources* (**a**) Internet, (**b**) photograph by N. T. Deręgowski, (**c**) Frame (2004), (**d**) Ethiopian Gospet, (**e**) Quian (2013)

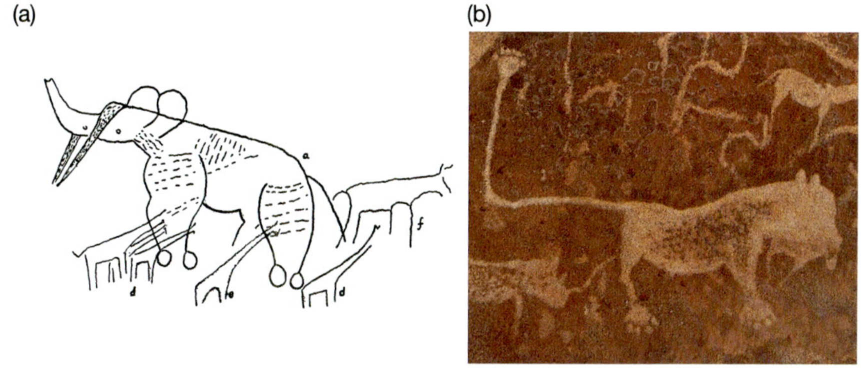

Fig. 1.14 **a** Tin Wana elephant displays his round pads each at the end of a leg portrayed by a central line (after Rodd 1938); the mammoths of the Chauvet cave depicted 300 centuries earlier show similar pads—an instance of remarkable stylistic stability. **b** The Namibian Tyfelfontein Lion waves a greeting with a paw at the tip of his tail. Other paws dangle at the ends of its legs, as do the pads of the Tin Wana elephant

EV: No, the head is not wrong.
AA: Is its belly wrong?
EV: No, it is not wrong, it is as it should be.
AA: Is his tail wrong?
EV: No, not the tail.
AA: So although, as they say, two wrongs do not make it right, three rights make it wrong!

When members of a distinctly different culture, Aberdonian school children, were asked to draw balsa wood models shown in Fig. 1.17. They drew their typical contours. When the same children were presented with a figure of an animal made with a flat horizontal body between a narrow vertical head and a vertical tail they overwhelmingly drew the animal as having all three segments in the same plane, twisting the figure just as Australian Aborigines twisted the crocodile.[18] This finding does not imply that Australian Aborigines saw the world in a childish manner, any more than Cezanne's paintings show that he did so when painting a coffee pot on the overburdened table (Fig. 4.14) but show that driven by a universal perceptual impulse they chose to depict its features thus.

A much-favoured model, more widespread in time and clime than the crocodile is the bovine. Its horned head presents the artist with difficulties of the kind presented by crocodiles—the typical contour of its head is in a plane orthogonal to that of its crown of horns.

A trio comprising ***a*** Palaeolithic artist of Alpera, a hieratic artist of ancient Egypt and a Bushman of southern Africa, congruously imposed coplanarity upon two elements whose typical contours are spatially orthogonal (Fig. 1.18). The perceptual impulse experienced by them was identical with that experienced by the Australian artist of Australia.

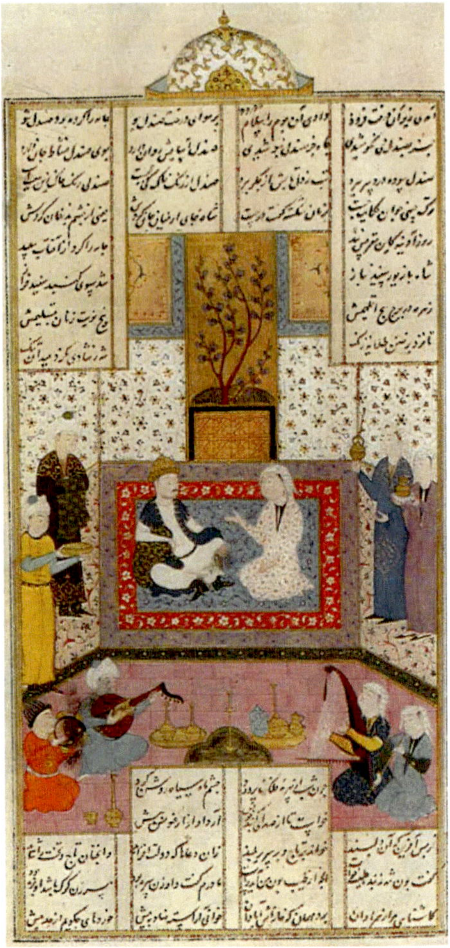

Fig. 1.15 A physician and his patient, both seated precariously on a rectangular rug (Moscow Anthropological Museum)

We have therefore a key to the rules of depiction governing that particular style, the fundamental of which is: whatever the orientation of typical contours in space their depictions should not be distorted since they represent highly significant visual features. The twisting by the Australian artists of the concatenated segments of a crocodile does not distort their shape, their entirely independent typical contours are preserved. (One would expect the effect of the torque on the abutting elements of a model (such as the Kakadu crocodile's body and tail) to depend on the extent to which they are seen as separate entities.)

Opportunities for such transformations are therefore probably more frequent in the world of nature than in the world of geometry. They disappear entirely whenever more than two non-coplanar surfaces share contours, as in the cube—each of its faces

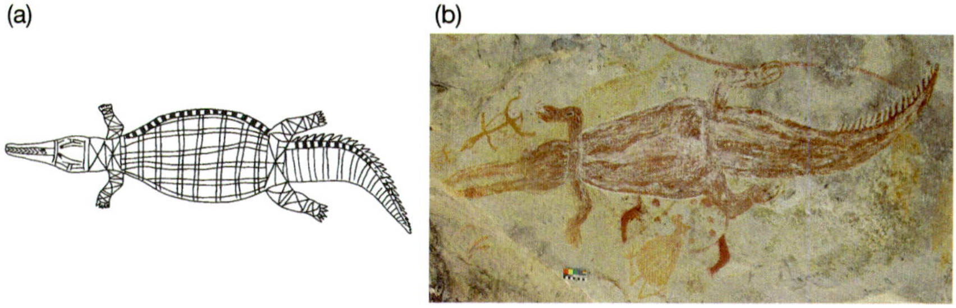

Fig. 1.16 Two Australian pictures of crocodiles, called here Kakadu Crocodiles for short. A drawing by Chaloupka (1984) and a photograph showing an in situ beast, kindly provided by Gunn et al. (2018), courtesy of Margaret Katherin

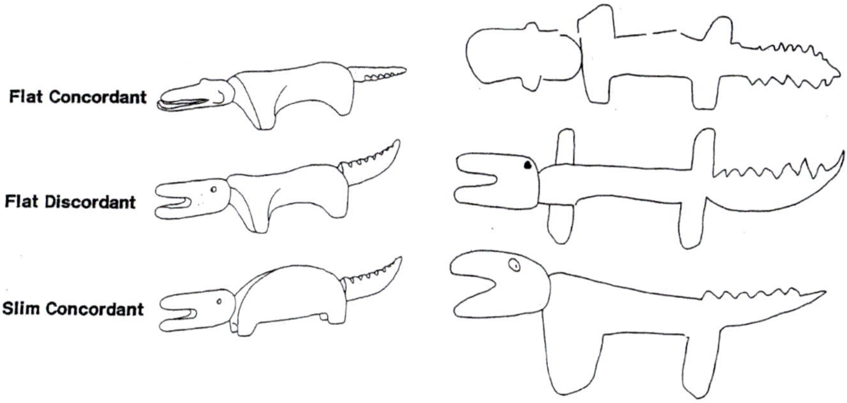

Fig. 1.17 Balsa wood models presented to Aberdonian children and their drawings of these models. *Top row model*: All elements have their typical contours in the horizontal plane. *Middle row model*: The head and the tail typical contours are in the vertical plane; the trunk's is in the horizontal plane. *Bottom row model*: Typical contours of all three elements are in the vertical plane. The middle row model is a spatial facsimile of an Australian crocodile and was responded to in the manner of Australian artists (Dziurawiec and Deręgowski 1992)

sharing contours with four other faces. Each face of a cube presents a square typical contour. Each side of this contour is also an element of a contour of an abutting face, and at each corner contours of three faces meet. An artist can draw a single square typical contour to represent a single face, or draw two abutting squares to show two faces sharing a segment of a contour, but cannot apply the Kakadu technique to show a third face if the squareness of its typical contour is to be maintained and shared contours correctly shown.

Everybody sees typical contours of objects but not everyone is equally expert at drawing them. When agricultural labourers unskilled in drawing, were required to draw a rectangular box they drew rectangular outlines, sometimes singly, sometimes in pairs a smaller rectangle within the bigger (Fig. 1.19). Concatenated rectangles

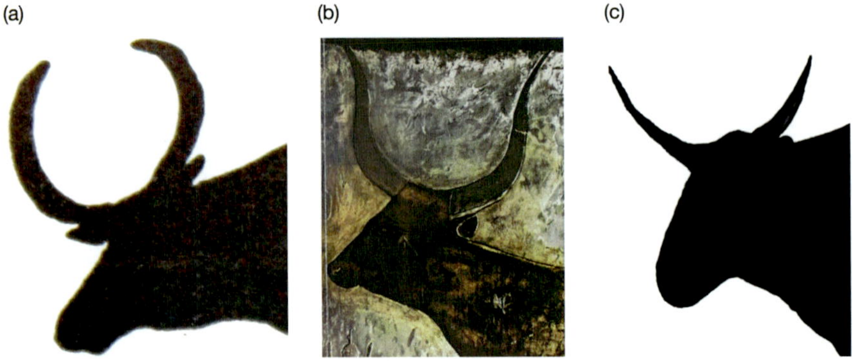

Fig. 1.18 Silhouettes of three cows in order of seniority: **a** Palaeolithic from Spain; **b** Ancient Egyptian; **c** Bushman, from South Africa. All three wear their crowns of horns so as to show their typical contours (after Swindler (1931), Michałowski (1969) and Tongue (1909)). Compare these figures with that of oxen by a Song dynasty painter Li Di (Fig. 5.5), which shows the power of foreshortening

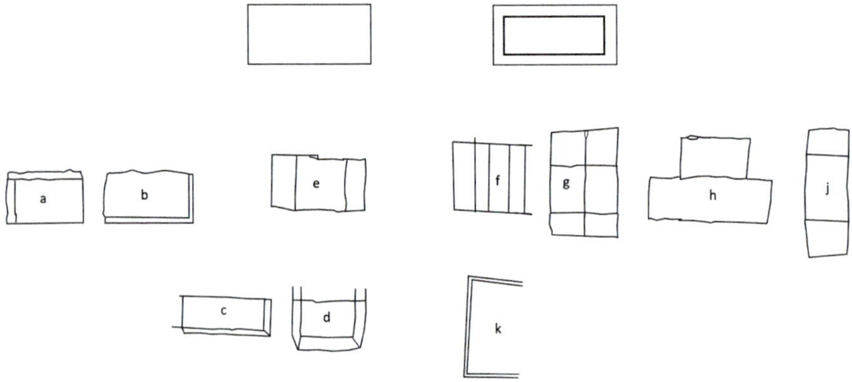

Fig. 1.19 Drawings of a rectangular box by illiterate men with no experience of drawing, obtained by Bartel (1958). Top row: a rectangle and concentric rectangles (81% of drawings). Middle row: concatenated rectangles (13%). Bottom row: three 'other' responses (6%)

were drawn by seven men. The remaining three figures are difficult to interpret. There was no pronounced tendency to abandon rectangularity of the outlines—typical contours are robust. Overall the drawings show that rectangular typical contours were readily abstracted and drawn. The men fumbled, however, at the task of putting together contours of different sides of the box.[19]

The prowess of typical contours is also made apparent by their effect on perceptually ambiguous items. Outlines of Bushman paintings are shown in Fig. 1.20. On the left (a) one of these is explicit, the other indefinite—it could be a cloud, a clump of tumbleweed, a sleeping bag. On the right (b) bringing them together to occupy

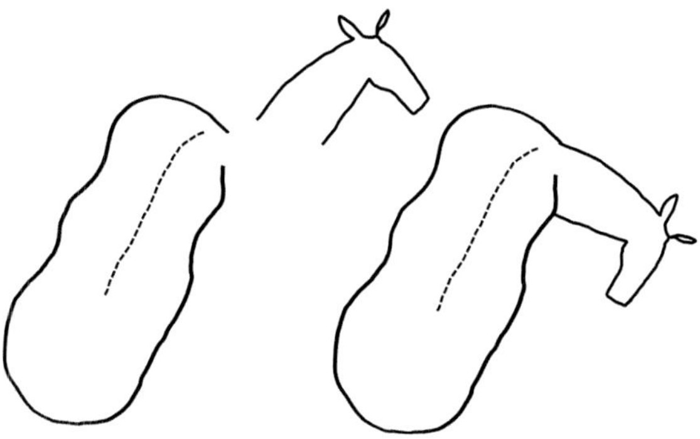

Fig. 1.20 A Bushman painting of an eland dismembered and reassembled (using outlines of the figure presented by Townley Johnson (1979))

positions which they had in the original painting converts the cloud into an eland's body. The conversion is perceptually unquestionable.[20]

A backward-looking eland presents contours which could be got by folding a paper cut-out of its typical contour. Pictures of such folded outlines do not question the value of the typical contour, but present its perceptually legitimate way of expanding artistic possibilities. In the bottom left corner of Fig. 1.21 the Bushman artist used this device to show an eland looking back. In the topmost portrayal Fig. 1.21 the device was pressed further. The picture is a puzzle. A similar puzzle but from an entirely different school of art is shown in Fig. 1.22, in which the entwined hands are clearly recognisable, but what about the rest? What does the picture show?[21]

Folding of typical contours can puzzle the viewer. Their perceptual unfolding may sometimes shock. Bronisław Malinowski[22] describes battles waged on the Trobriand Islands thus: "Rows of men faced each other, armed and holding wooden shields. The shields of the bravest of the brave had designs painted on them. Such a shield was a challenge. Splitting it was a heroic act." Such a design is shown in Fig. 1.23a. To the uninitiated it is an entirely innocuous pattern. It remains obscure until perceptually unfolded. Unfolding (Fig. 1.23b) reveals its true nature. It is a picture of a vicious witch with huge protuberant pudenda. Such witches are know to fly at night and under the guise of pleasure infect men with elephantiasis. The shield screams at the enemies in front: "May she visit you! May you get it!" The change of the significant contours into typical contours reveals the truth.[23] Surprising as the result of unfolding may appear, unfolding itself is a simple matter. It is like full opening of a rudimentary pocket knife with blades hinged at opposite ends.

The very skilled artist may also modify the typical contour by slow degrees so as to depict the process of metamorphosis, such as takes place when a man turns into a frog. A record of such an event is shown in Fig. 1.24. It was obtained by an observant Bushman. It is a telling document showing that inverse of the Darwinian theory can be entertained. The record of the metamorphosis shows that it progresses by abandoning the typical contour in favour of a foreshortened stance in which the man becomes more frog-like, and that this process continues until he becomes a frog. (The argument that the document should be seen as showing metamorphosis of the frog into the man is dismissed on the evidence of oral tradition.) Figures from a Peruvian graveyard[24] also show transformations, such as that of a man changing

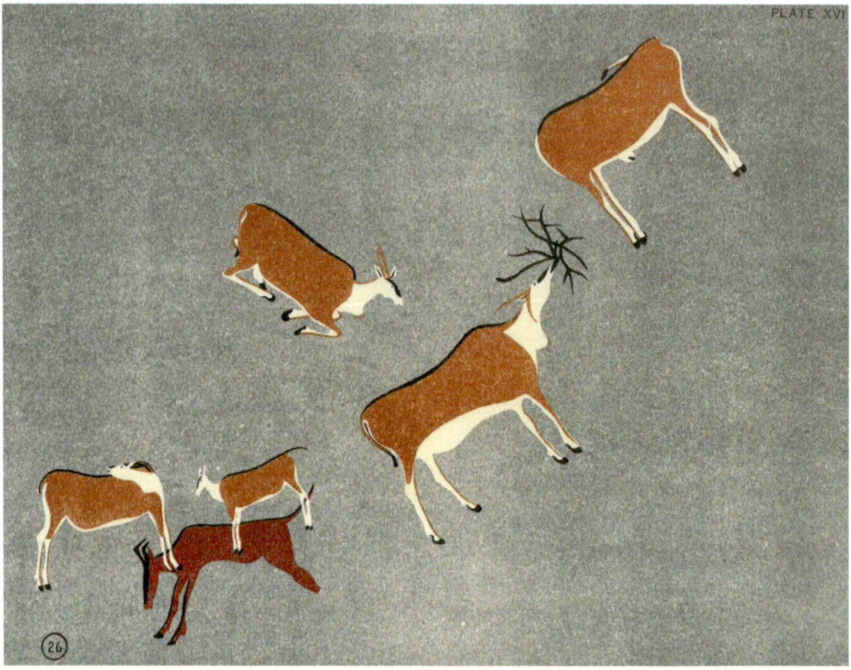

Fig. 1.21 A Bushman painting of eland. The eland shown on the extreme left displays its folded typical contour. That on the extreme right presents an insoluble puzzle. Not quite. The uplifted tail identifies its rear end of the animal, its snout peeps below its belly (after Tongue 1909, Plate XVI)

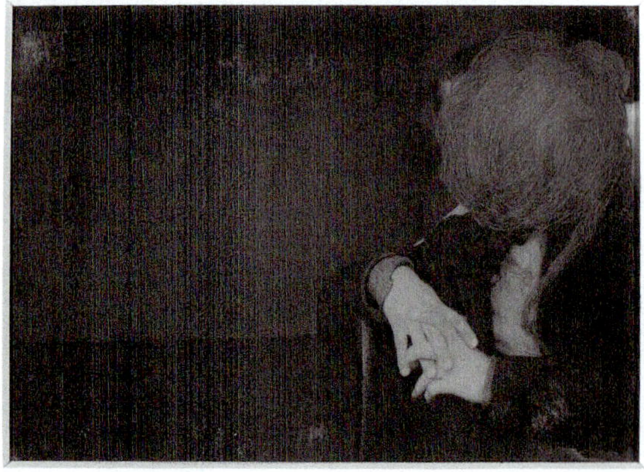

Fig. 1.22 Jakimowicz presents a puzzle, as did the Bushman artist. It is entitled "A kiss". The title hints at the correct percept[21] (Courtesy of Muzeum Narodowe, Warszawa)

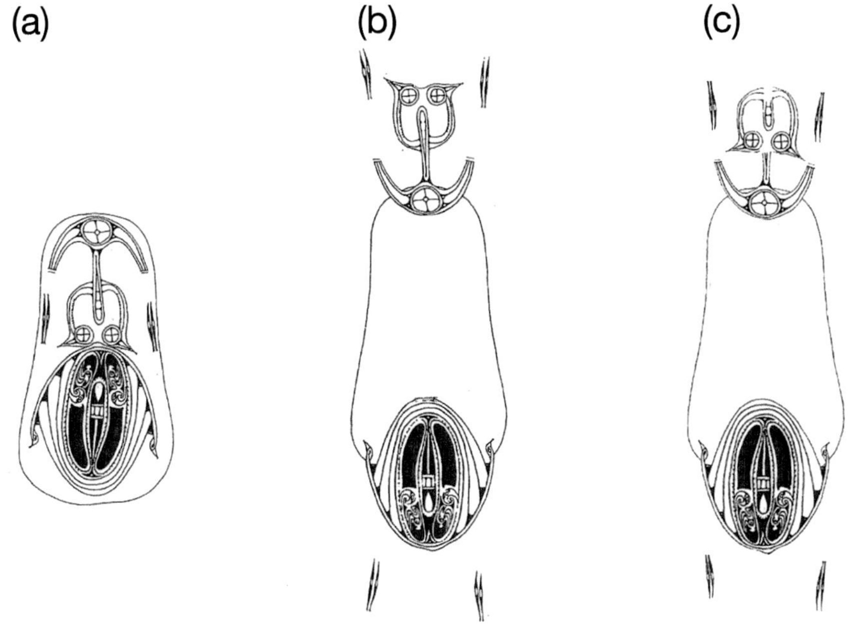

Fig. 1.23 A Trobriand shield pattern (**a**) and its two transformations: **b** simple unfolding of the head and the pudenda as suggested by Bushman art, **c** much more complex transformation suggested by Leach, in which some elements play a dual role (eyes/breasts) and the head remains in its original orientation after unfolding. The seminal notion that transformation may elucidate the shield pattern was put forward by Leach (1954). Leach uses the term folding-up to describe the transformation he has in mind, but the figure appended does not show such a transformation. The orientation of the folded-up head, for example, is the same as that of the untransformed figure. Simple unfolding implies (see Fig. 4.11) that b becomes d (unfolding about a vertical axis) or that b becomes p (unfolding about a horizontal axis); when b becomes q the perceptual process is either that of double unfolding b $\rightarrow$ d (vertically) and then d $\rightarrow$ q (horizontally), or that of rotating the image in the picture's plane through 180°, much more complex procedures

into a shark, but these transformations are of a less subtle kind, they consist essentially of exchange of parts, their origin remaining identifiable—their typical contours are not affected. Thus as Fig. 1.25 shows, the unmistakable shark's head and tail combine with equally unmistakable human legs and arms. There is no subtlety of changing contour characteristic of Bushman art. In Peruvian art the contours are not affected. They are retained to preserve the identifying attributes of the individual parts. Escher's magical transformation in his Sky and Water, of fish into birds (or, perhaps, birds into fish), comes closer to the Bushman transformation, for it involves gradual emergence of typical contours from an undefined background. Emergence of a new contour is not, however, as subtle as is the gradual change of an existing contour.

Typical contours are assertive whatever their origin and are therefore assertive when defining eyes. The spatial setting of human eyes differs from that of bovids. Human eyes are so set that their typical contours lie in the plane orthogonal to the

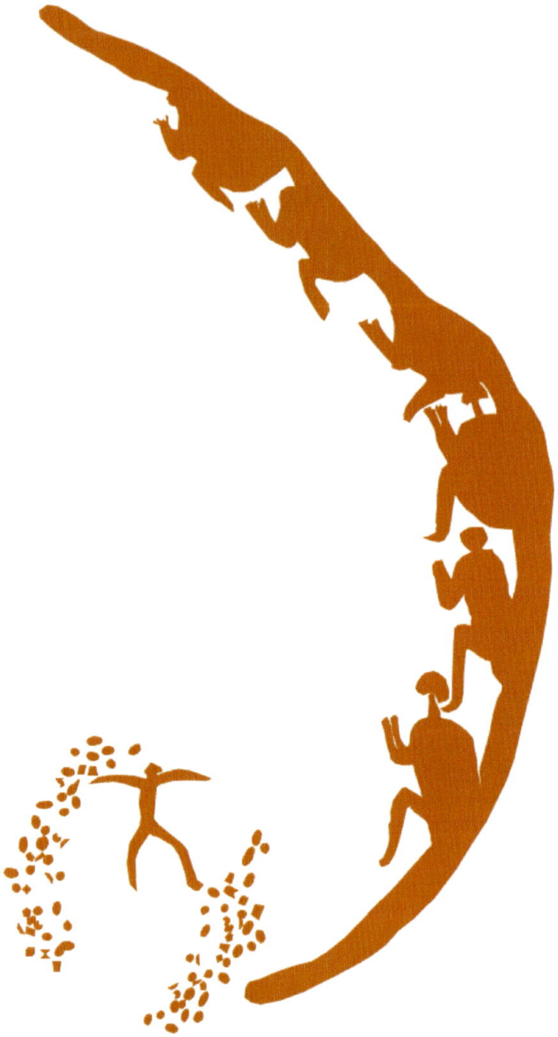

Fig. 1.24 Bushmen record of gradual metamorphosis of a man into a frog (Tongue 1909)

plane of human profiles. Bovids' eyes are within the plane of their typical contours. The artist depicting a man or a woman encounters the problem which Zeuxis Primus, the artist depicting a bovid, is spared.

The problem of the former is that of presenting, by typical contours on the picture's surface, two entities which are mutually orthogonal. It can be solved by adopting the manner practiced by the Australian painters of Kakadu crocodiles, or by abandoning typicality of one of the contours.

The former approach yields a profile with an eye that gazes at the viewer, a characteristic of Ancient Egyptian, Cretan (Fig. 1.26),[25] and other cognate art styles. Styles which compound typical contours within each other, and do not differ in

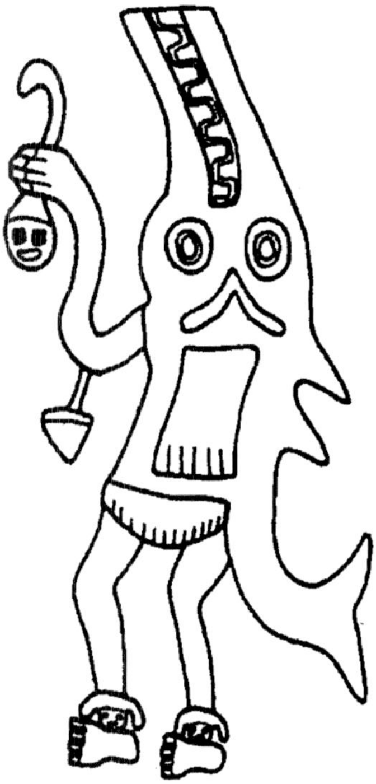

Fig. 1.25 Somewhat less convincing metamorphosis than that observed by the Bushman artist of man into a shark recorded in Peru (note that the feet show typical contours)

essence from that of the Tyfelfontein lion and Kakadu crocodile. This style which must have seemed to the Greek vase painters of the sixth century BC somewhat unsatisfactory. They abandoned it by small degrees over a period of a century and a half—such is the intransigence of a typical contour. The contour of the eye in the final picture of this lengthy transformation, shows no traces of typicality (Fig. 1.27).

A different fate befell drawings done by children. Over a century ago children's drawings of people had, as a standard form, the two eyed profiles (Fig. 1.28). Typical contours, the profile of the face and those of the eyes, were rendered as co-planar. The historical contest in the nurseries of the West between these two typical contours was won by the eyes, an outcome contrary to that reached by the Ancient Greek vase makers.[26] In portraiture of bovids the typical contours of their bodies dominate. The typical contours of the eye are but embellishments which may or may not be placed within the main contours.

In the Bushman art of South Africa eyes of animals are only sometimes shown. The Palaeolithic animals of Altamira and Lascaux do not show eyes as a rule and the positioning of presumed eyes is often uncertain. Eyes are neglected because the contribution that a depicted eye makes to understanding of many a picture is negligible; a chase, a dance, a battle are unmistakably conveyed by the gestures of the participants, and gestures are conveyed by the central lines and by typical contours. When eyes are depicted in the art forms just mentioned they do not differ

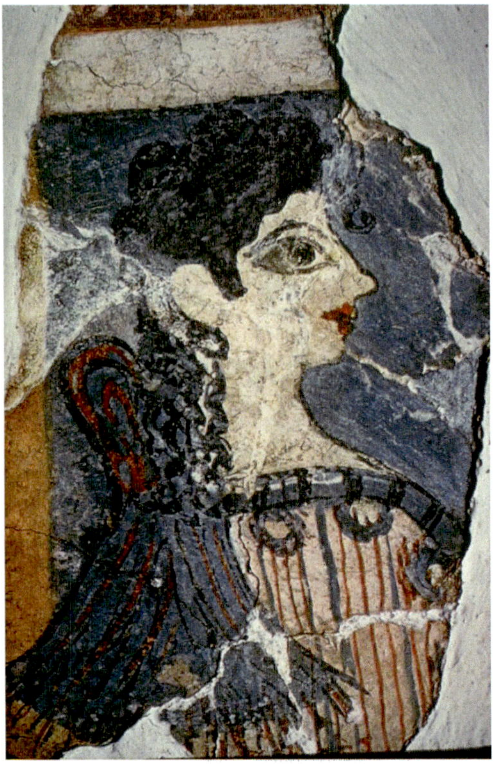

Fig. 1.26 A head and an eye both shown in their typical contours in a Knossos fresco. The guiding technique is the same as that of Australian artist; typical contours of the eye and the profile are unchangeable (after Swindler 1931)

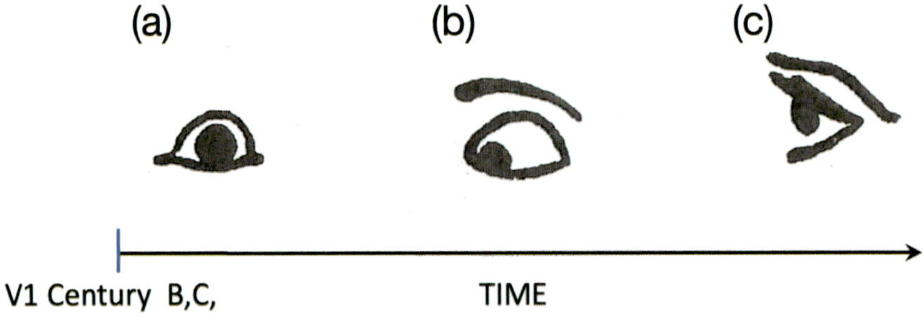

Fig. 1.27 The story of the gradual change of the portrayal of the eye in the head rendered in profile, on the Grecian pottery (after Swindler 1931). **a** Sixth century B.C. portrayal showing typical contour. **b** Gradual deterioration of the typical contour. **c** After 150 years no trace of the typical contour remained[19]

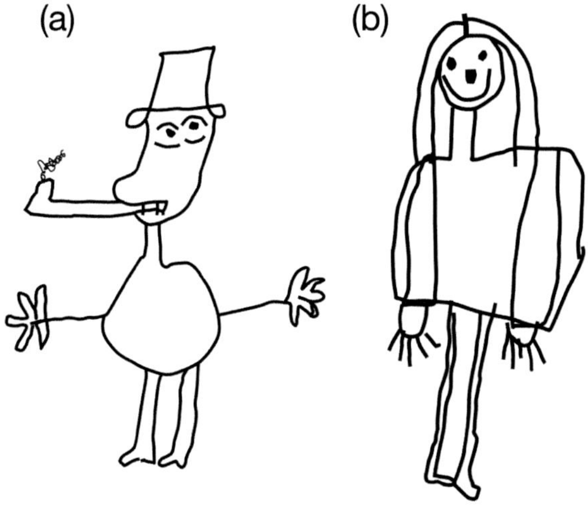

Fig. 1.28 Children's drawings **a** two-eyed profile prevalent in the nineteenth century,[26] **b** frontal view of today by Miss Halszka Le Riche

systematically among the species. The reason for portraying eyes seems therefore to be completeness of the picture, which is attained by placement of either their typical contours or their substitutes, sometimes mere dots within the peripheral contour.

Notes

1. References to animal camouflage and mimicry quoted, unless otherwise stated, derive from Cott (1966).
2. See Bursill (1859/60). It is a remarkable compendium of ingenious designs for "hand shadows to be thrown upon the wall".

   The relationship between shadows cast by objects and depictions of objects has a long history. Pliny the Elder (23?–79) speculating on the origin of depictions (Naturalis Historiae, Lib. XXXV) notes that Greeks (perhaps those of Sicyon, perhaps those of Corinth) are said to have discovered the art of drawing and that they arrived at the discovery by tracing outlines of shadows cast by men (XXXV; V). It is a charming story, which is accompanied by another anecdote about cast shadows and art. When the lover of the beautiful (one assumes) daughter of Butades, the potter of Sicyon, was about to set out on a journey abroad, she traced the outline of the shadow of the beloved head thrown on a wall by a lamp. A painting by George Romney (*The Origin of Painting*; ca.1775–1780) and by Jospeh-Benoit Suvée (*The Invention of Drawing;* ca. 1791) portray this event (see Petherbridge, 2010). This tracing so impressed her father that he pressed clay onto the wall and made a relief which he fired together with his other pottery (XXXV; XLIII). A new art was thus born.

   Shadows cast by randomly formed solids, e.g. by coagulations of molten wax poured on water, are used in some cultures, to divine the future. Such a shadow simplifies the percept by confining relevant information to the peripheral contour. This phenomenon is acknowledged by the cognoscenti of the shadow casting art, the puppet makers for Balinese shadow theatres. Their puppets' feet project typical contours (see Fig. 1.13a).
3. Barrière (1976) presents figures which in his view show that fingers were amputated.
4. The Rorschach open-ended, projective personality test consists of a set of random but symmetrical ink blots. Those are presented, one at a time, and subjects are required to say what

they see. The responses reveal, as one would expect, considerable inter-subject differences. Different people do see different pictures in ink blots. However, interpretation of the variation of individual responses is difficult. The test is not widely used nowadays (see Semeonoff 1987).

5. See Mach (1897). The Necker cube (Fig. 1.3c) is an equally well-known reversible figure, first described by Necker (1832). Now the upper, now the lower square face is seen by the viewer as being nearer. Both the Mach's book and the Necker cube reverse as if they were solid objects. Solid objects also can reverse. For example, a hollow cone viewed monocularly reverses and is seen as a tepee, so that objects sliding down its inner surface appear to move towards the viewer (Deręgowski and Tatler 2016). It is noteworthy that after inversion the perceived change of size of the moving object is contrary to that expected—objects seen by the viewer as approaching are also seen as decreasing in size, a fact showing that is necessary to maintain the integrity of the perception. For discussion of Mach Bands and their influence on vision and art, see Marmor and Ravin (2009).
6. von Fieand and Moustgaard (1977, p. 359).
7. The figures are taken from Breuil and Obermeier (1935). Extensive references to this seminal work of archaeology will be made. An animal presenting its spinal typical contour is either stationary or moving in the observer's fronto-parallel plane. When so moving it encounters least environmental resistance (of wind, water, flora). The environmental pressure fosters symmetry about the plane—asymmetrical items would tend to veer. Such symmetry is characteristic of many creatures (quadrupeds, fish etc.) and not surprisingly it holds for many vehicles (locomotives, bicycles, etc.)
8. The picture is in the manuscript collection of the British Library: MS Royal 16 G.v.f. 54v. XV cent. French. To analyse the effect, cover the bottom half of the picture and note that in the case of both upright frames the left-hand member appears closer to the viewer. Now cover the top half of the picture and the left-hand member of the left-hand frame appears to be further away. A transparency of the picture, cut in two horizontally at about the level of the spinner's elbow, and one half rotated relative to the other removes the perceptual confusion. The same remedy can be applied to M.C.Escher's well known Belvedere. Such perceptual inconsistencies are not always noticed. Children may recognise two elements of a drawing as representing a solid and fail to see that such a solid could not exist (Young and Deręgowski 1981).
9. Reverse conversion of a cube into a hexagon is easy to imagine, but when more complex figures are concerned imagination fails. It is not easy to suggest significant contours which would convert the "Old Growler" (Fig. 1.1) into a pair of hands.
10. Gibson (1950) drew particular attention to the effect of the "gradients of stimulation" upon perception of depth embracing under this heading not only monocular texture gradients, but also binocular gradients, "… retinal disparity, like other signs of depth, could be defined as gradient of stimulation—not of the single retina, it is true, but a gradient of the theoretically combined images of the two retinae (p. 72)." It appears (see 5 above) that abandonment of one half of such a binocular texture gradient leads to perceptual inversion.
11. Fortes (1940).
12. This scarcity is pointed out by Leo (1977). Hardi and Saaghy (1981) confirm it. (They also report on the basis of their study that a stick figure "expresses the abstract geometric tendencies of schizophrenics", a topic outwith the scope of the present essay.)
13. Rodd (1938).
14. Ramsayer (1977), for a description of Balinese puppet theatre.
15. Frame (2004). The embroideries depict composite stylised creatures (Fig. 1.13b). The "Swimmers" among them have their eyes "shown from above as if the viewer looks down on them in the watery realm, but the mouth, gills, body and human arm are shown in profile." A schema similar in principle to that of Australian artists drawing crocodiles. A figure on a funerary cloth (Fig. 1.13c) derived from a photograph taken by T. Deręgowski shows feet with four toes, just like the uplifted hands.
16. Qian (2013).

17. This lion is unique. There appear to be no other similar portrayals, a surprising fact in the art which repeatedly uses similar motives.
18. Dziurawiec and Deręgowski (1992).
19. See Bartel (1958). A brief report of his study was presented by Deręgowski (1986). Bartel speculates that Fig. 1.19 showing concentric rectangles may have been intended to show thickness of the walls of the box. This supposition is not entirely convincing as the box was closed when displayed and no attention was drawn to the thickness of its walls.
20. The original painting is now in the Natal Museum.
21. The Bushman figures are from a collection of tracings published by Tongue (1909). The impact of this publication, originating in such unexpected place- sub-Saharan Africa, was considerable. Rodger Fry, a greatly influential art critic of the day, wrote that "nothing could be more unlike than some of these (copied) scenes and that the artists, when drawing silhouettes (of men and animals) show an alertness to accept the silhouette as a single whole instead of reconstructing it from separately apprehended parts" (Fry 1910). The Kakadu crocodile is an example of such reconstruction. The other picture, *A Kiss,* is by M. Jakimowicz. The entwined hands are of the seated man. A small white triangle of his forehead is visible near the top. The tangled head of hair belongs to the woman who is leaning from behind over the man's shoulder. The picture is in the National Museum, Warsaw. I am grateful to Dr. C. Creed for his observations on the picture.
22. Malinowski (1920).
23. The unfolded figure (see Fig. 1.23b) shown here is not that proposed by Leach (1954), whose paper introduced unfolding as an explanatory notion. Leach's figure (c) cannot be obtained by simple unfolding. The witch in his figure looks towards a distant horizon, whilst the simply unfolded figure looks towards pudenda. She is taking aim.
    One can easily conceive the following scene:
    "The medicine missionary says that it is carried by mosquitoes", said Jacob, who worked as a steward at the mission house.
    "This is strange, there are swarms and swarms of mosquitoes. Does each mosquito carry a little piece of the disease?" asked his brother.
    Then, the old man with long silvery beard, as silvery as mosquito wings, said very slowly, "There are many, many mosquitoes and everyone here is stung by many of them. We should all be sick."
    There was silence.
    Then Jacob, the steward, said, "The missionary is mistaken."
    "Not mistaken, but not fully informed," said the old man. "He thinks that all that look like mosquitoes are mosquitoes. Some poisonous berries look edible. Witches can easily change into mosquitoes and then change back. Some witches look like mosquitoes. The missionary does not know this, and if you tell him this he will not believe you."
24. Frame (2004).
25. Figure 1.26 is after Swindler (1931), who also provides a table of 30 examples of gradual mutation of the depicted Grecian eye. Figure 1.28 shows items 1, 17 and 29 of that table.
26. Nineteenth century child's drawing (from Sully, 1896) and a drawing by its twenty-first century counterpart (Halszka Le Riche, aged 7). Wilson and Wilson (1982) wittily examine this temporal change of style.

References

Barriere C (1976) L'art parietal de la Grotte de Gargas: Palaeolithic art in the Grotte de Gargas. Universite de Tolouse Le Miral Instituit d'art prehistorique, British Archaeological Reports, Oxford

Bartel K (1958) Perspektywa Malarska. Państwowe Wydawnictwo Naukowe, Warszawa

Breuil H, Obermeier H (1935) The cave of Altamira at Santilana del Mar, Spain. Topografia de Archivos, Madrid
Bursill H (1859) Hand shadows to be thrown upon the wall. Griffith & Farran, London
Chaloupka, G. (1984) From palaeoart to casual painting. Darwin Northern Territory Museum of Arts and Sciences
Cott HG (1966) Adaptive colorization in animals. Methuen & Co., London
Deręgowski JB (1986) Kazimierz Bartel's observations on drawings of children and illiterate adults. British Journal of Developmental Psychology 4: 331–333
Deręgowski JB, Tatler BW (2016) The eye of the illusion: Variation in perceptual phenomena within an inverted percept. Perception 46: 100–108
Di Leo, J.H. (1977) Child Development. Brunner-Mazel, New York
Dziurawiec S, Deręgowski JB (1992) Twisted perspective in young children's drawings. British Journal of Developmental Psychology 10: 35–49
Fortes, M. (1940) Children's drawings among the Tallensi. Africa, 13, 239–95
Fortes, M. (1981) Tallensi children's drawings, In Lloyd, B. and Gay, J. (Eds). Universals of Human Thought. Cambridge University Press, Cambridge
Frame M (2004) Motion pictures: symmetry as animator, classifier and syntax in Nasca embroideries of Peru. In: Washburn DK, Crowe DW (eds) Symmetry comes of age: the role of pattern in cultures. University of Washington Press, Seattle
Fry R (1910) Bushman paintings. The Burlington Magazine for Connoisseurs 16:334–338
Gibson JJ (1950) The perception of the visual world. The Riverside Press, Cambridge, Mass
Gunn, R.C., Douglas, L.C. & Whear R.L. (2018) The complexity of Arnhem Land rock art complexes. Rock Art Research, 35, 3–24
Hardi I, Saaghy M (1981) The Problem of Stick-figure. In The Pathology of non-verbal communication. V.Andreali (ed.) Masson Italia Editori, Milano
Leach ER (1954) Trobriand Medusa. Man, 54:103–105
Mach E (1897) Contributions to the analysis of the sensations. Open Court Publishing Company, Chicago
Malinowski B (1920) War and weapons among natives of Trobriand Islands. Man 20:1–12
Marmor MF, Ravin JG (2009) Artist's eyes: Vision and the History of Art. Abrams, New York
Michałowski K (1969) The Art of Ancient Egypt. Thames & Hudson, London
Necker, L.A. (1832) Observations on some remarkable phenomena seen in Switzerland and an optical phenomenon which occurs on viewing of a crystal or geometrical solid. Philosophical Magazine 3, (1), 329–37
Petherbridge D (2010) The Primacy of Drawing, Yale University Press, New Haven and London
Qian G (2013) The Huashan rock art site (China): the sacred meeting place for Sky, Water and Earth. Rock Art Research 30: 22–32
Ramsayer V (1977) The art and culture of Bali, Oxford University Press, Oxford
Rodd F (1938) Some rock drawings from Aïr in the Southern Sahara. Journal of the Royal Anthropological Institute of Great Britain and Ireland 68: 99–111
Semeonoff B (1987) Projective techniques. In: Gregory RL (ed.) The Oxford Companion to the Mind. Oxford University Press, Oxford
Sully J (1896) Studies of Childhood. D. Appleton and Company, New York
Swindler MH (1931) Ancient Painting. Yale University Press, New Haven
Tongue HM (1909) Bushman paintings. Clarendon Press, Oxford
Townley Johnson, R. (1979) (T. Maggs, ed.) Major Rock Paintings of Southern Africa. David Philip, Cape Town
Von Fieandt, K. Moustgaard (1977) The Perceptual World. Academic Press, London
Wilson M, Wilson BG (1982). The case of the disappearing two-eyed profile. Or how little children influence the drawings of little children. Review of Research in Visual Arts Education, 8: 1–8
Young A.W. & Deręgowski, J.B. (1981) Learning to see the impossible. Perception, 10, 95–105

Chapter 2
On Perception of Depth, Real and Portrayed

Such is the perceptual system's eagerness to detect meaningful entities that it detects typical contours, even in accidental patterns. Given the viewer's goodwill, all kinds of creatures and objects will be seen in an inkblot; segments of its periphery deemed to be similar to segments of typical contours of objects and creatures. Even when such goodwill is lacking silhouettes showing typical contours, such as those commonly used on warning signs, are instantly understood. In all such pictures typical contours epitomise the depicted objects. They convey what the objects are without indicating their girth. Such pictures are not mistaken for the objects they portray, They have a word-like quality.

The eye which seeks a meaning everywhere, even in inherently meaningless patterns such as inkblots, has yet another tendency—that of seeing depth in patterns on flat surfaces. The perceived illusory depth is quite distinct from the implicit depth resulting from recognition of an object. Recognition of a silhouette of an elephant as an elephant is not associated perceptually with perception of the animal's girth, although it does not deny its corpulence. The illusion of depth influences the perceived spatial relations among elements of a picture. Whilst it is tempting to describe the performance of someone trying to make sense of an inkblot as similar to that of a drowning man grasping anything within reach and therefore yielding varied and disparate responses among individual observers (an occurrence pleasing to the student of individual differences, but not to the student intent on finding perceptual regularities), the reason for occurrence of any particular response is elusive; this is not so in the case of perception of illusory depth. Here it is commonly held that such percepts are related directly to percepts encountered daily in the real 'three-dimensional' world. If our daily experience, for example, that the more distant elements of our environment are seen less clearly than those nearby, this experience translated, for example, to the picture showing two squares, one black, the other light grey (see Fig. 2.1) implies that the darker (the more salient) square should be seen as relatively nearer to the viewer. This is generally so. Viewers see the black square as nearer to them than the grey square. (They also see the grey square as the

J. B. Deręgowski, *On Picture Making and Picture Seeing*,
Vision, Illusion and Perception 4, https://doi.org/10.1007/978-3-031-23348-7_2

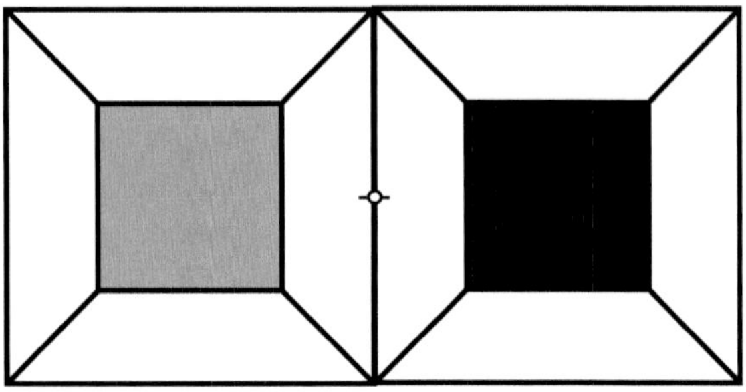

Fig. 2.1 Effect of salience on perceived distance from the observer. The black square is generally seen as nearer than the grey square. Since perceived stimulus size is affected by the perceived distance (size constancy)—the grey square is generally seen as larger of the two

larger of the two. This is another story. The story is that as the squares are equal and project equal images into the eye, the square further away must be really larger.)

The two mutually independent tendencies, that of seeing meaningful figures and that of seeing depth differ in the nature of their cognitive involvement. The tendency to see a figure is hesitant, its fruits are uncertain and can be modified at will. The tendency to see depth is direct and stable; when a change of perceived depth occurs it generally does so spontaneously, not driven by the observer's cognitive set.

These two tendencies are at the root of picture making. How can the artist evoke illusion of depth of the inclined surface of a depicted object?

The solution is to present within the peripheral contour, cues mimicking depth cues of the real world. Not all real-world depth cues are adaptable to the purpose. Since the picture is drawn on a flat surface the entire cohort of depth cues resulting from binocular vision is absent. The monocular cues that can be used are those of clarity of vision. Clarity is affected by two factors, the angle which a model sustains at the eye and spatial obfuscation. Both these factors cause clarity of the model to decline with the increase of its distance from the observer; the angle decreases and the presence of only apparently innocuous elements, such as particles of dust or smoke in the intervening space increases. The view enjoyed by farmer McDonald leaning on his gate is obfuscated both by the smoke of his pipe and by mist; and the gate to his neighbour's distant field projects onto his eyes a smaller image than an insolent rabbit who is eating his cabbage, a few yards away. Farmer McDonald looks lovingly on the neat rows of his cabbages. The rows are seen less and less clearly as they stretch into the distance; somewhere in the distance the rows appear to merge. The same happens to the grass growing on the narrow baulk separating cabbages from carrots. They all appear to converge to the same distant point, but they do not do so in the same manner. Farmer McDonald sees only the tops of the cabbages in the row directly in front, but he sees both tops and sides of those in rows to the left and right. Indeed, the greater the lateral displacement the more he sees of their sides. Farmer McDonald sees the texture gradients. He does not think of the view in these terms, nor does he reflect on the fact that their translation onto a picture's surface would create an illusion of depth and that they harbour the seeds of linear perspective.

Farmer McDonald's urban cousin, Doddie Munro, looking fixedly straight ahead along a level straight street sees lamp posts and kerbstones not only *appearing* smaller and less distinct with increasing distance, but also sees them as converging with distance. The lamps do so along lines sloping downwards and the kerbstones do so along upward sloping lines. (The effect was described by Euclid (fl. 300 BC), who did not have the benefit of long, paved and illuminated streets [1]). A nearer object may appear either higher (as lamps do) or lower (as kerbstones do) than a more distant object of the same ilk. The latter experience is markedly more frequent—we look on the ground more often than we look at the sky. The cue is clearly ambiguous, but the ambiguity is not even handed. Notwithstanding this ambiguity, a sloping line does evoke a fundamental perceptual effect of depth. A sloping line may derive either from the kerbstones or from overhead lamps. Either of its ends may represent the most distant or the nearest end of the depicted object to the viewer, but in both guises, it retains its perceptual link with depth and the artist wishing to create an illusion of depth in the plane of his picture has this resource at his disposal. Oblique lines are agents of pictorial depth.

An observant observer walking past a wall on which a straight line inclined to the horizontal has been painted will, if he steadily looks at the oblique line, notice that it appears to move. A laboratory confirmation of the phenomenon can be obtained using a pointer, hinged at one end in such a way that it can be set in any direction.

In Vermeer's picture "*The Music Lesson*" the meeting of the floor and the receding wall defines a sloping line. When the picture is looked at at close quarters, from a different point on the line parallel to its surface and the pointer set to parallel the direction of the sloping line, setting changes markedly with the observer's stance. The sloping line is not seen as an element fixed within the picture's surface, but as a line in space. Similar, but somewhat less pronounced changes of perceived orientation are seen in a purely geometrical diagram derived from the picture (see Fig. 2.2). The changes are less pronounced because the diagram lacks depth cues such as the linear perspective of the windows, the texture gradient of the floor, present in the picture[2].

Other works of art evoke similar percepts. Two paintings by J. Chełmoński[3], a nineteenth century realist painter, can serve as examples. In his *Four in Hand* (Fig. 2.3a, b) the four horses harnessed to a carriage rush with breakneck speed at the viewer wherever he may stand. If the viewer moves to the right, the horses sway to the right still intent on trampling him. If the viewer moves to the left the horses sway to the left. Perceived horses' headings from two different stances are shown in figure b. In "The Gossamer", a much more tranquil scene, a young cowgirl lies on the grass playing with gossamer (Fig. 2.4a). Her right arm points upwards, her left is flat on the ground. A viewer walking past this picture sees the girl's left arm swerve. This illusory movement would also occur if the girl in the picture were replaced by an arrangement of central lines of her body, limbs and the cudgel; if she were depicted by means of central lines. The resulting arrangement of the central lines, so created, would form Visual Illusion figures involving perception of depth (Fig. 2.4b). (Angles between the line of the outstretched arm and two parallel lines, that of the body and that of the cudgel, are like the dominant angles of isometric drawings discussed in the next chapters).

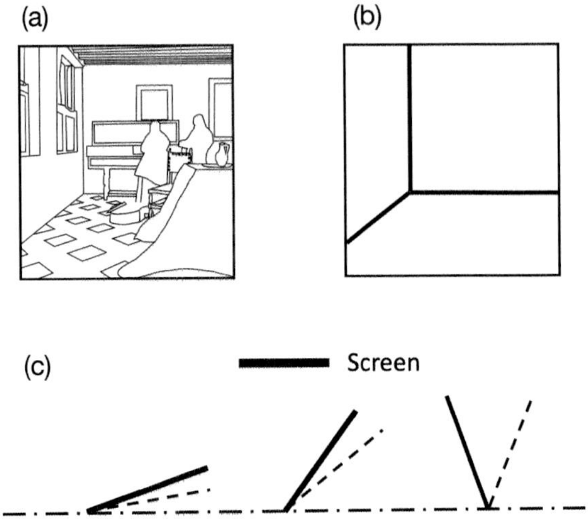

Fig. 2.2 (**a**) An outline of the essential features of Vermeer's *The Music Lesson*. In the reported experiment a colour slide of the picture was used (**b**) Three key lines derived from wall and floor joints in Vermeer's picture (**c**) Angular settings of the pointer to show the direction of the sloping line as seen by the observer standing in front of the picture to the left or to the right. Continuous lines show responses made to Vermeer's picture, dotted lines responses to the geometrical diagram (**b**)

A visitor to the Cracow museum where Chełmoński's Four in Hand picture is displayed may notice, not only that the four horses are intent on trampling him, but also that eyes of various portraits follow him wherever he goes. Marmor (23) offers the following analysis of the phenomenon. When one moves from a stance in front of the nose of a sculpted head the optical projections of various features change; shapes of both the nose and the earlobes become more clearly defined, but when viewing a *picture* of the sculpted head taken enface under identical conditions, there are no comparable changes, hence since the observer has moved but its perceptual relationship with the model has not changed the model must have moved accordingly.

The observed effect is striking. Social Psychologists would say, because eyes are such important means of social communication. The said Social Psychologist may even be tempted to quote, inebriated, Ben Jonson: "Drink to me only with thine eyes, And I will pledge with mine …".

The eyes, or rather eyespots, therefore would seem to provide a basis for construction of percepts just as doess the obliquity of the central lines of Chełmoński's horses.

Visual illusions have been hunted and pursued by psychologists with great vigour ever since their very existence was recognised in the nineteenth century[4] because a true theory of perception must account for perception of all the stimuli including visual illusions. The Horizontal-Vertical illusion (Fig. 2.5) is the most venerable member of the clan. Two equal linear segments, one vertical, the other horizontal, are perceived as grossly unequal—the vertical seems longer. The reason for the effect is obscure, but speculation that its basic universality is moderated by an individual's experience has been widely entertained. Projections on the retina of the two linear segments of the illusion figure, it is postulated, are seen in the same manner as any other projections, and perceptual similarities of projections affect percepts. Any line extending away from the eye projects a foreshortened image and the extent to which the viewer moderates the foreshortening is influenced by experience.

"If one lives in an environment that provides many opportunities for looking at horizontal expanses, one should become subject to the tendency to infer long, frontal plane horizontal distances from short, vertical retinal images"[5]. Therefore, one would expect a population living in such an environment to be more susceptible to the Horizontal-Vertical Illusion, as measured by the ratio of lengths of the

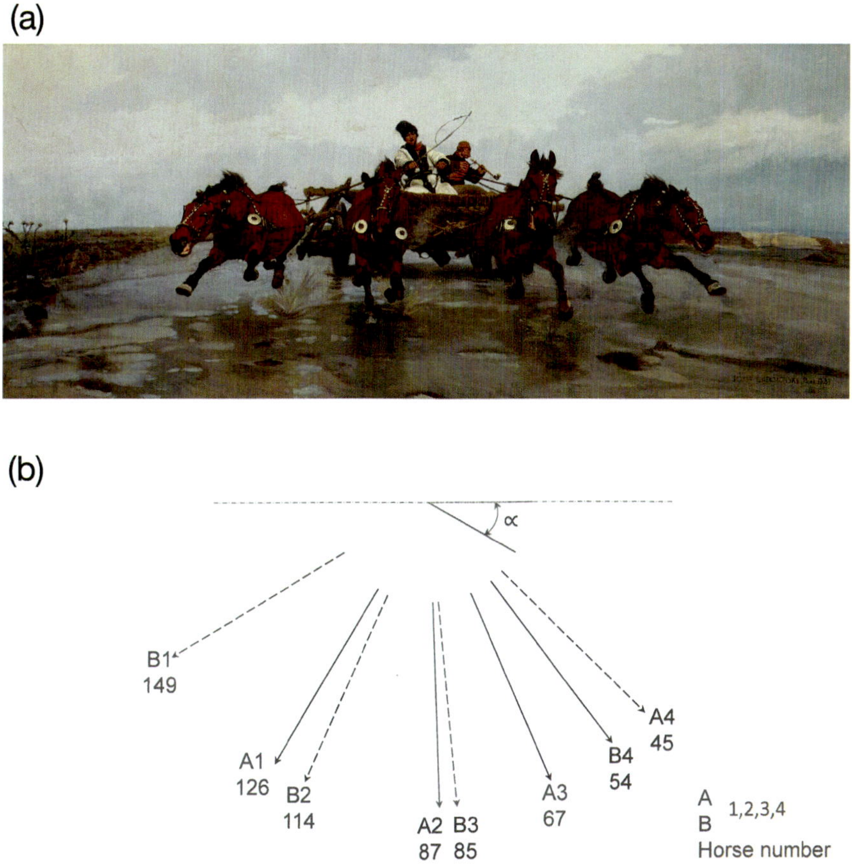

Fig. 2.3 *Four in Hand*, by J. Chełmoński. (Courtesy Muzeum Narodowe Kraków). The picture is notable for the number of oblique features: The drivers hat and whip, the postillion's horn, and especially for the obliquities of the central lines of individual horses. The latter obliquities ensure that as the viewer's stance changes, so do the perceived directions in which the horses are heading. Extensions of the elements of the harness, if drawn, would converge on the driver's chest with the stock of his whip establishing the point of convergence. Lines drawn from the point of convergence to suggest edges of a road would increase pictorial depth markedly. The picture shows a roadless steppe. It was not painted on the steppe but in a Paris studio. Attempt to paint such a picture on the steppe using Leonardo's pane, the horses charging towards the painter, would probably be the last attempt by the artist to make any picture The diagram shows the means of empirically obtained perceived headings of the horses (1–4) as seen from two stances, that in front of the middle of the picture (**A**) and that removed 7.5 m to the left (**B**)

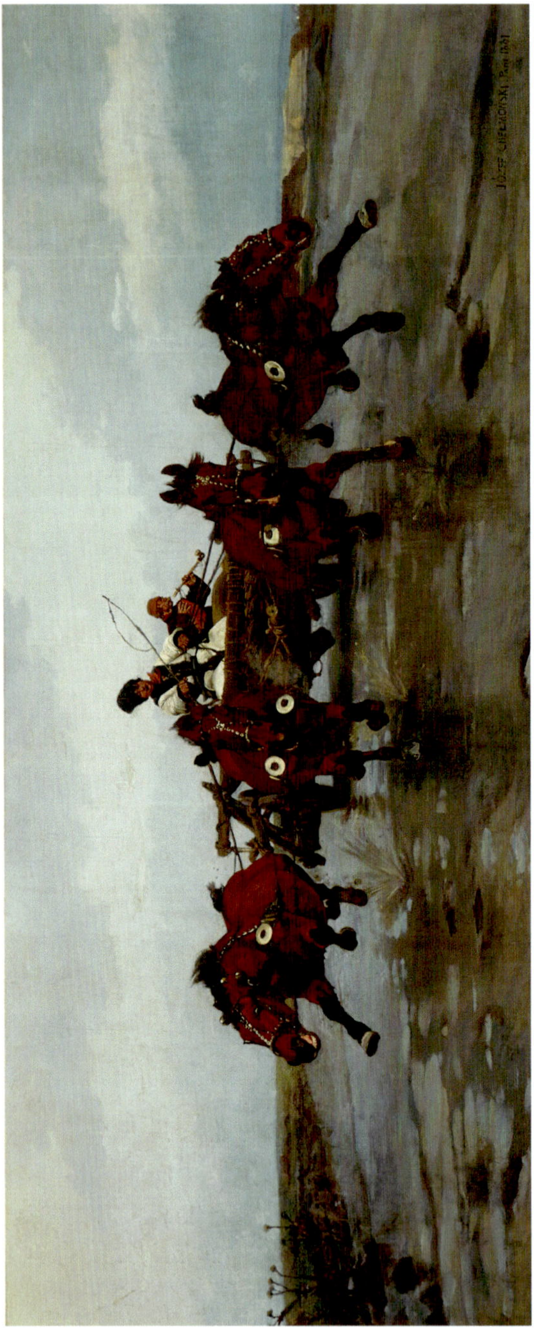

Fig. 2.3 (continued)

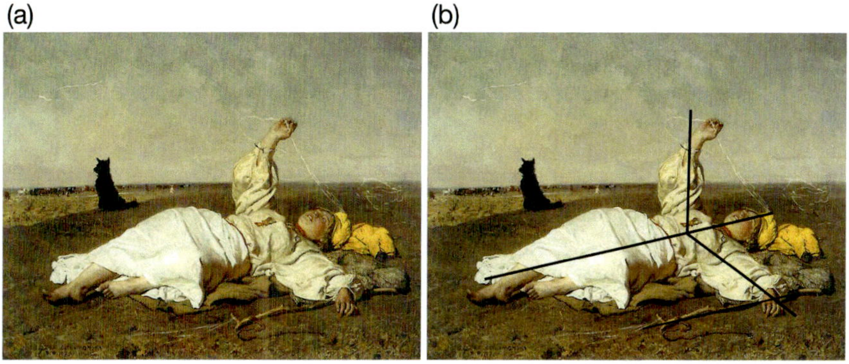

Fig. 2.4. **a** *Gossamer,* by J. Chełmoński. (Courtesy of Muzeum Narodowe, Warszawa). The body of a girl offers central lines which are much more perceptible than are those of individual horses, in Fig. 2.3. Identified, they form a pattern of four straight segments, one of which is vertical, as shown in Fig. 2.4 **b**. The angle between the central line of the body and that of the left arm changes with the viewer's movement. (Inserting the central line of the dog would strengthen the illusory effects.)

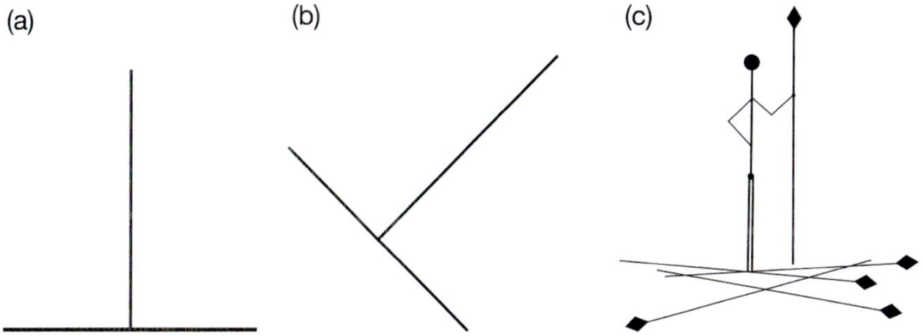

Fig. 2.5. **a** Horizontal-Vertical illusion figure used by M. H. Segall, D. T. Campbell and M. H. Herskovits (1966). The T-type figure was a rotated T shape with two elements separated by a small gap and differing in colours (black and red). **b** The effect of rotation of the T-type illusion figure through 45° **c** Use of the illusion in portrayal. The spear chosen by the warrior is not the longest of the set, All spears are of about the same size

two segments, than a population whose environment does not offer such opportunities. A large body of cross-cultural data was gathered to investigate this hypothesis using stimuli in orientations shown in Fig. 2.5a and was judged to provide a "good fit", but for a few qualifications. One such qualification is revealed by rotating the stimuli through 45°. The hypothesis implies that, when both segments of the rotated illusion figure are equally, perceptually, foreshortened the illusion should not occur. Figure 2.5b shows the illusion figures turned through 45° and therefore projecting equally foreshortened segments. Yet in the case of the T-shaped figure the illusion, albeit weakened, blatantly persists. This gratifyingly complies with Swift's[6] observation on the laws of nature:

"So, naturalists observe, a flea.

Hath smaller fleas that on him prey;

And these have smaller fleas to bite'em,

And so proceed ad infinitum, …".

The artist is free to exploit the illusion. It is so exploited in Fig. 2.5c. The warrior's pride in his chosen spear is unjustified. It is not the longest spear of the four, all spears are drawn of the same length. The elongation, that of the vertical spear, is the result of the horizontal-vertical illusion.

Another well known illusion, the Muller-Lyer, is more explicitly pictorial (Fig. 2.6). The shafts and the arrowheads of each of the two figures form patterns readily seen as those formed by typical contours of rectangular solids (as shown by couplings a-a', b-b'). This similarity suggests the hypothesis that familiarity with rectangular solids increases effectiveness of the illusion, the shafts, which are physically equal, are not seen as such—the shaft of Fig. 2.6a is seen as longer than that of Fig. 2.6b. This effect is likely to be moderated by the extent of the observer's experience of such visual patterns. The greater the frequency with which objects such as a' and b' are seen, the greater the proneness to the illusion. Since such objects are found in different frequencies in different cultures, cultures provide means of testing the implied hypothesis. Inhabitants of cultures in which right-angularity is infrequent, houses are not rectangular but hemispherical (as are e.g. those of the traditional rural Zulus) the shape of the cultivated plots and pathways is determined by geographical contours, rather than by land surveyors' rules; there are but few pieces of rectangular furniture, the hypothesis suggests, likely to be less prone to the illusion than the inhabitants of industrialised societies in which such right-angled and rectilinear objects are commonplace, whose cultures are therefore called carpentered. (Elmslie 1901), a missionary and a physician living among the Ngoni, describes such a culture. "The tents of the people are built in circles around the cattle fold. Like everything the native makes they are circular, and he points to the sun, moon, and horizon as a reason why they should be so." An attempt to evaluate this hypothesis using a large cross-cultural sample was not entirely successful. The hypothesis was not supported as wholeheartedly as one would desire, most notably because children (whose cultural experience is of necessity less than that of adults) were not consistently less prone to the illusion than adults. An unambiguous and striking relationship between Muller-Lyer illusion figures and objects in space (Fig. 2.6)—is immediately apparent to members of carpentered cultures. In cultures lacking carpentered models such strong relationships cannot be established, but although weakened, the illusion is experienced there, showing its non-cultural roots[7]. Such roots are exposed by the fact that, although a carpentered environment can project only fins of the Muller-Lyer figure, with the inter-fin angle within a 90°—180° span, the illusion is experienced when the inter-fin angle falls outwith this span. This is shown in Fig. 2.7. The middle column shows illusion figures having the inter-fin angles larger than the right angles; figures of the first column have the inter-fin angles of 90°, those of the third column

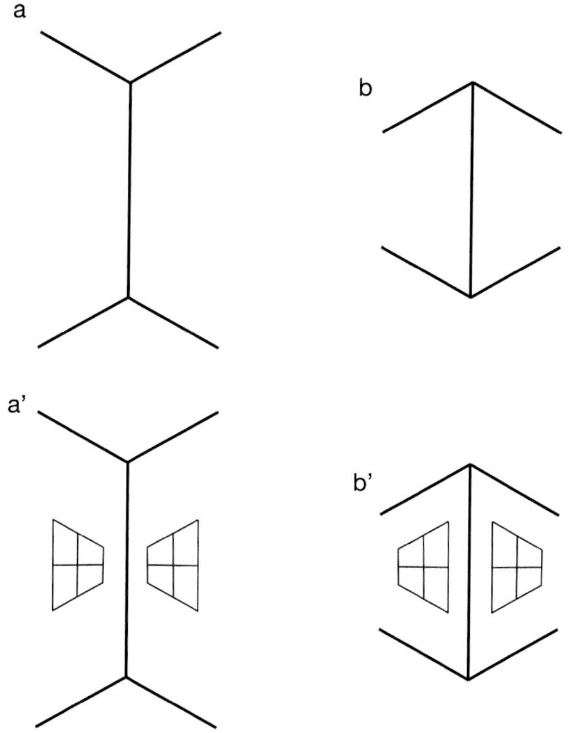

Fig. 2.6 Muller-Lyer Illusion figures. Top row figures in their basic form, bottom row two of their supposed environmental relations found in the world where right angles are often seen; the carpentered world

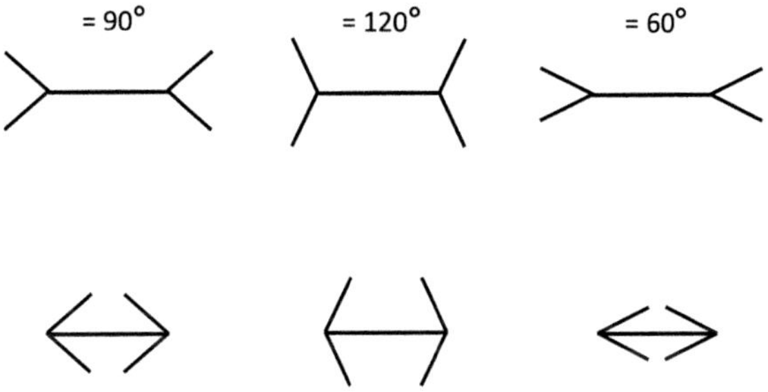

Fig. 2.7 Influence of the inter-fin angle on illusion. First column angle of 90°; middle column: angle of 120°; third column angle of 60°

have the inter-fin angles smaller than right angles. In all three pairings the stem of the top figure appears to be longer: The illusion is seen.

At the core of the Muller-Lyer illusion figure and its derivatives is the oblique line. An oblique line evokes vague but potent perception of pictorial depth. The evocation is vague because it lacks definite sense, as shown in Fig. 2.8 of a corridor. The contours formed by intersection of walls with the ceiling and floor are labelled a, b, c, and d, and are identical in pairs a-c and b-d, although within each pair one of the contours has its upper end close to the door, while the other has its upper end close to the viewer. The sense in which each contour points is not conveyed by the

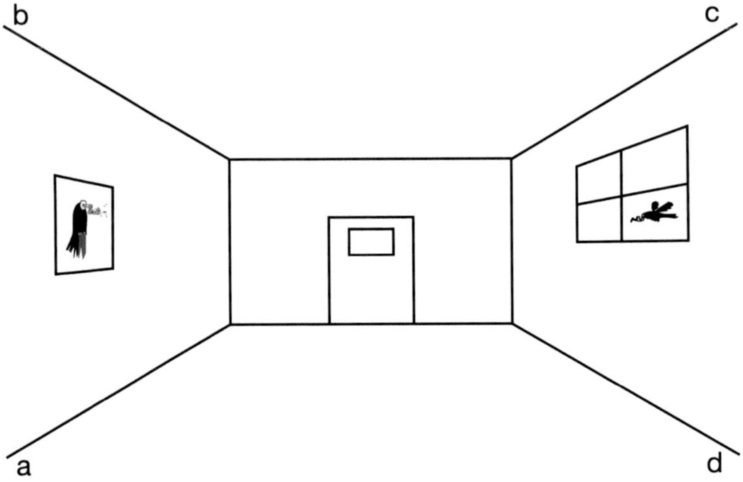

Fig. 2.8 The door at the corridor's end. Obliquity does not of itself indicate which end of the oblique line is closer to the viewer. The figure could also be of a helicopter landing pad at the top of a pyramid

contour itself, but by ambient cues. A simple investigation shows the importance of the cue of obliquity. It was conducted on a population whose contact with pictures was very limited. Two wire models and two drawings, shown in Fig. 2.9, were used in the investigation.. The wire models were made of stout wire, each consisting of two identical squares and a connecting rod. In model *a* the squares were in parallel planes and the connecting rod was normal to them—the model which looked like an incomplete skeleton of a box was clearly spatial. In model *b* the squares and the connecting rod were all in the same plane—the model was flat. The drawings were conventional representations of the two models: Drawing *a*' of model *a* and drawing *b*' of model *b*.

The subjects were either presented with the spatial model *a* and both drawings and asked which of the pair was the better picture of the model, or they were presented with one of the pictures and both wire models, and asked which model did the picture show.

The choices made were overwhelmingly as follows. In response to model a, drawing b' was chosen. In response to drawing b', model b was chosen. In response to drawing a', model a was chosen. The relationship between models and their depictions is clearly asymmetrical. The reasons for asymmetry are:

1. The spatial model *a* presents two distinct squares and so does drawing b'; drawing a' does not do so.
2. Model *a* is, unlike model b, spatial and drawing a' has an oblique element which embodies a hint of space; model *b* is flat.

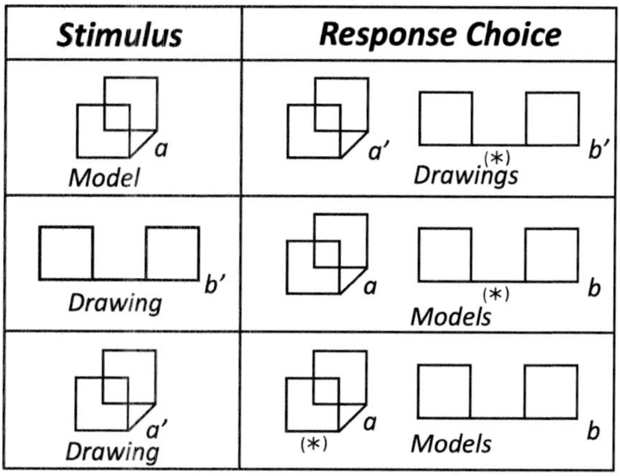

Fig. 2.9 Stimuli and response choices in a simple preference task performed by people of little familiarity with pictures. Subjects responded to a three-dimensional wire model *a* by choosing one of two drawings *a'* or *b'*. They responded to drawing *b'* by choosing one of two wire models: Three-dimensional *a* or flat *b*. They responded to drawing *a'* by choosing between the wire models *a* and *b*. Asterisks show dominant choices made

The result obtained does not imply that those making choices were aware of this spatial hint of obliquity, any more than those responding to such illusions as Muller-Lyer are. Knowing artists exploit this fact; unknowing viewers of their pictures are affected.

The Bushmen who decorated a shelter near Washbank (South Africa) were, one suspects, such knowing artists. In the picture (Fig. 2.10) two Bushmen and a dog are out on a hunt. The legs of the Bushmen form obliques which contrast with the horizontal line formed by the dog's legs. Of this picture R.E. Fry, the art critic, wrote: "… here is the silhouette of a most complicated gesture with foreshortening of one thigh and crossing of the arm holding the bow over the torso rendered with apparent certainty and striking verisimilitude"[8].

The vagueness of the sense of an oblique segment is not clarified by addition of other equally vague elements. Convergence of three linear segments evokes pictorial depth but the percept is ambiguous, the point of convergence may be seen by the viewer as either close by or far away, as the reversals of Mach's Book demonstrate[9]. Ambiguity lays the percept open to a variety of influences. The happy family shown in Fig. 2.11 is, according to reports of Scottish viewers, gathered inside a rather sparsely furnished room with rather a small window. This is not what their rural African counterparts reported. For them the family was in the open. The comely young woman has just returned from the river and was gracefully balancing a four-gallon tin full of water on her head. A window or a four gallon can? Convergence of the three lines within the rectangular boundary does not give a clear answer but indicates pictorial depth in both cases. One very seldom encounters such three convergent lines in their skeletal form, in perfect isolation, without any hint whence the light is coming. This hint is conveyed by shadows.

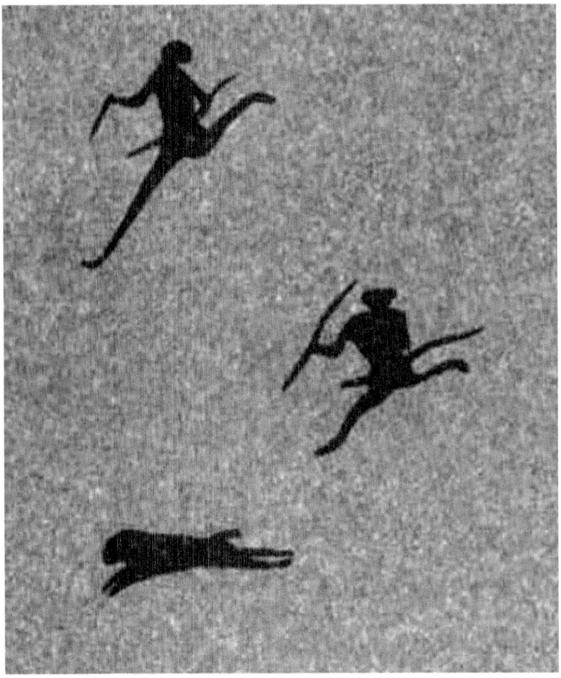

Fig. 2.10 When Bushmen go a-hunting (after a rock art tracing by Tongue, 1909). The obliquity of the hunters' legs contrasts with the horizontality of the dog and converts uniform rock face into space

Light and shadows: Zeuxis Primus is not interested in light. He wants to depict a real object. He wants to portray a bison in the cave where the only light present will be that of smoky cressets, but the kind of bison he remembers has been defined by shadows.

There are three kinds of shadow: (1) Shadows within the peripheral contour cast by some features within it upon other, such as folds of cloth, thus revealing some of the characteristics of the surface; (this shadow referred to in books on art as *modelling* has been in painters' repertoire continuously since times immemorial. (2) Shadows cast by models on abutting surfaces which remain visually attached to the models and so tend to form single visual entities. Such enlarged entities are more noticeable, so when conspicuousness is not desired an attempt is made to reduce the shadow. Some butterflies use this subterfuge and when resting with folded wings align them with the sun's rays. The artist may find a shadow cast by an object just as irksome as butterflies do, not because he wishes his picture to be inconspicuous, but because he does not wish the portrayal to be marred by depictions of irrelevant, ephemeral appendages. Finally, there are shadows which are detached from the model and so constitute visually separate entities. Their perceptual independence from their origin is the essence of Mr Bursill's shadow casting act. Such independence makes it possible for the shadow of a hand to become that of a goat.

Zeuxis Primus has, of course, seen, as has everyone else, all three types of shadow, but to him, only the first kind seems marginally relevant. The pendant shadow of the second kind is merely confusing. It is not an enduring part of the casting object to

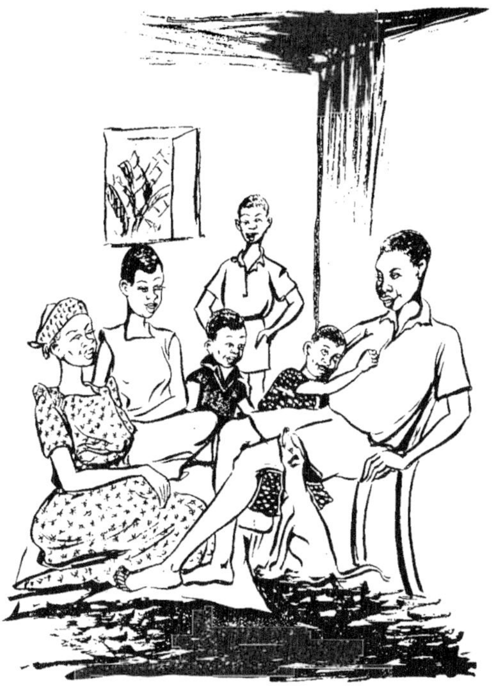

Fig. 2.11 A happy family (with permission of UNESCO, 1963). The greater the ambiguity of pictorial expression the greater the role of culture in determining the percept. Is there a four gallon can on top of the young woman's head, or is she just seated below a window?

which it is attached; it is an ever-changing appendage. The third kind, the detached, cast shadows whose shape depends on the orientation of the surface on which they are cast and on which they are often compounded with shadows of other objects, may (as Mr Bursill has shown) not even relate to the models casting them in a visually meaningful way. The shadows of the first kind which can be found within the typical contour of the bison fill in the contour so that the surface within it differs from the background.

Palaeolithic paintings of animals, in the Altamira Cave in Spain, present a puzzle[10]. Most depicted animals poise like ballet dancers, *sur les pointes,* on the tips of their hooves, as the bison in Fig. 1.5 shows. Such prancing is not frequently observed in mammals. A possible solution to the puzzle is as follows.

Imagine a low stool standing in front of you (Fig. 2.12A). The ends of the pair of legs closer to you (*a*) would appear in its picture below those of the *b* pair (Fig. 2.12C). Now imagine the same stool tipped over so that the *a* pair of legs is above the *b* pair (Fig. 2.12B), which lies on the floor, The picture changes: The ends of the *a* pair are now above those of the *b* pair (Fig. 2.12D). (In all figures the eye is co-planar with collateral legs *a* and *b*.) One can therefore deduce from the relative elevation of ends of the legs in a picture whether the stool is upright or on its side. What holds for stools holds for bisons. Spatial orientation of Altamira bisons[10] can be deduced from the relative elevation of their hooves.

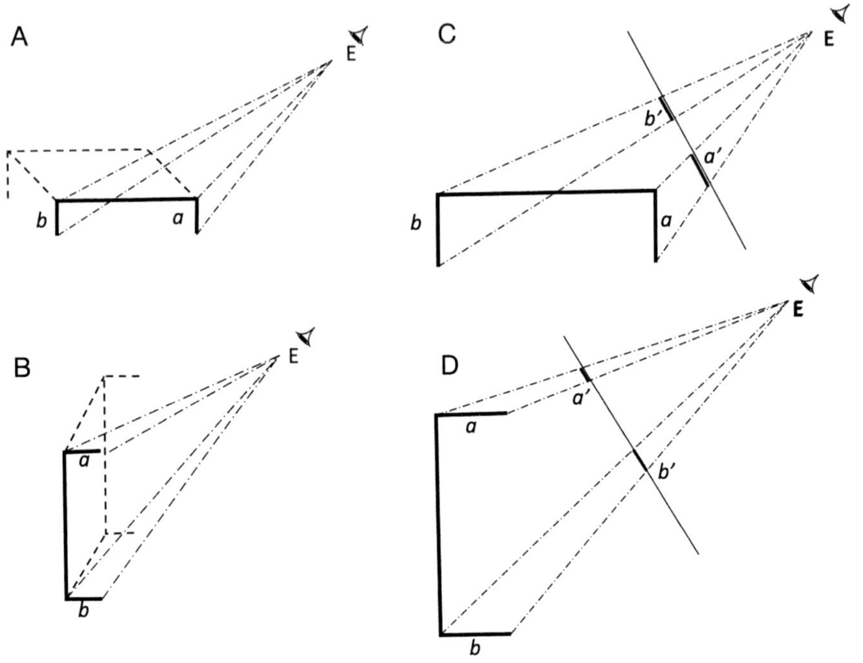

Fig. 2.12 Two views of a low stool as seen from above by the eye co-planar with the legs a and b, and, below, its projections onto pictorial plane. When it is upright the near legs (**a**) appear in the picture's plane below the more distant legs (**b**). When it is tipped over the relationship is reversed and the legs nearer to the viewer (**a**) appear above those further away (**b**)

Figure 2.13 shows four modified versions of Fig. 1.5; the bisons' legs have been selectively shortened. In figure (a) legs of the pair seemingly further from the viewer have been shortened so as to create the impression that the animal is indeed standing. In figure (b) legs of the pair seemingly closer to the viewer have been shortened to create the impression that the animal is not upright. In figure (c) the forelegs derived from figure (a) have been combined with the hind legs of figure (b), thus creating a 'twisted' animal, analogous effect is presented by figure (d) which combines forelegs of figure (b) with rear legs of figure (a), but the beast is twisted in the opposite sense.

Allocation of the "upright" Altamira bisons to the categories thus defined yields no upright bisons, infrequent bisons lying on their flanks and a numerically dominant group of contorted bisons[11]. This suggests that the portrayals are of dead animals. Such a view agrees with Leason's[12] assertion that *sur les pointes* portrayals are portrayals of animals' legs after rigor mortis, when relaxed muscles allow hooves to droop. This drooping does not, however, account for the depicted rotated hooves. Thus, for example, the bison's body in Fig. 1.5 is shown in side view. In relation to the body the hooves are rotated clockwise, and show approximations to their typical contours. The artist, it appears, did not wish to present typical contours in their

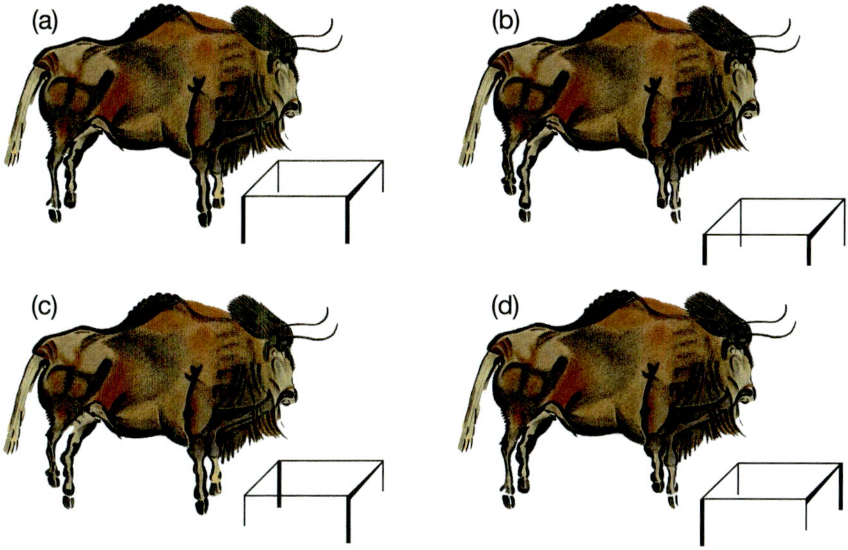

Fig. 2.13 Four modifications of the bison of Fig. 1.5. A sketch of a glass-topped table having analogous attributes is juxtaposed with each bison (**a**) Perceptually the nearer right legs are *lower* in the picture; the bison is standing (**b**) Perceptually the nearer right legs are *higher* in the picture; the bison is on its side (**c**) Perceptually the nearer right foreleg is *lower* than the left in the picture, and the perceptually nearer right hind leg is *higher* than the left in the picture; the bison's body is twisted (**d**) The perceptually nearer right hind leg is lower in the picture, and the perceptually nearer right foreleg is higher in the picture; the bison is twisted

fullness, as they were presented by those who drew hooves and feet discussed in the first chapter, but did drift towards typicality.

Bisons' legs that dispose of the notion that the depicted animals were alive, dispose also of the more general notion that "any art which has not been affected by adoption in Greece around 500 BC of a method of representation that takes note of perspective, oblique views and foreshortening" does not embrace the notion of pictorial depth[13].

The assurance with which it can be said that a picture embodies pictorial depth depends on the presence of pictorial depth cues. Presence of a single depth cue bears no convincing evidence that the artist intended to evoke pictorial depth, as it might have arisen accidentally. When two or more cues jointly indicate a characteristic with a frequency greater than that expected by chance, it is reasonable to hypothecise that they may share an origin. When pictorial depth cues do so it is reasonable to assume that the artist was aware of their significance, and used them to create an illusion of depth.

A pair of bison's legs (Fig. 2.14, see also Fig. 1.5a) presents two depth cues: The overlap cue, the leg nearer to the viewer overlapping the more distant; and the size cue, the claws nearer to the viewer being larger. In the figure shown these cues are in agreement; the overlapping leg has bigger claws. Evaluation of the portraits of

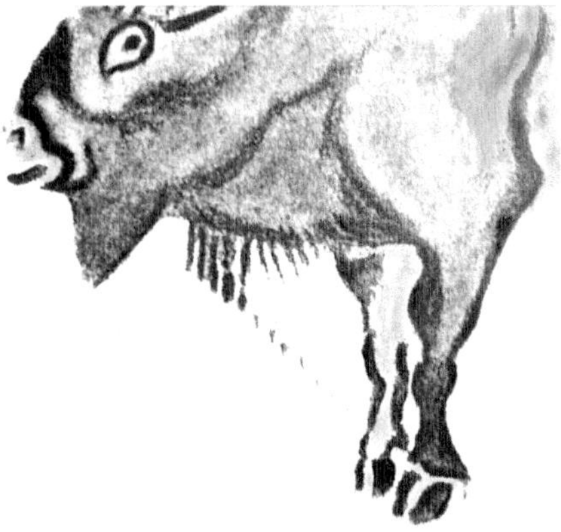

Fig. 2.14 Depicted bisons' hooves decrease in size with increasing distance from the viewer. This questions the theory that pictorial space was not seen until discovered by Ancient Greeks

bisons in the Altamira cave in terms of these attributes shows that such consistency of the two cues prevails with a frequency that is unlikely to be a result of chance[14].

Most Altamira bisons with undistorted typical contour of the spine and therefore seemingly erect show concord among the depth cues; it follows that Palaeolithic artists were aware of pictorial depth. This finding questions the popular notion of the Greek innovation and suggests that fifth century Greeks did not discover previously unknown pictorial depth cues, but that they merely threw off relics of Egyptian hieratic rigour. Pictorial depth cues are of no importance, the Egyptian artist would argue. Typical contours aided by certain well-established conventions ensure perfection in art, ensure that all is well[15].

Use of typical contours advocated by the Egyptian artist is not an entirely effective way of portrayal. All is well as long as the portrayal is of the familiar. Lack of familiarity does not, however, hinder viewers from interpreting pictures and may let their imagination to roam in imaginary spaces. This is most likely to happen when the yen for confabulation overrides the perceptual reality. Nothing is more satisfactory than a good yarn and of all good yarns, gory yarns with supernatural flavour are the best; and so on a wall of the Lascaux cave a battle between shamans is seen by many. One of the shamans, in the guise of a bird-headed man, lies prostrate on the ground. His adversary is in the guise of a bison. The bison is wounded. Its bowels are spilling out. Their cow's udder shaped contours dangle below its belly (Fig. 2.15a).

A parietal painting by Bushmen in the Ndedema Valley[16] shows a herd of eland heading east. From the belly of one of them, peacefully walking, dangles a similar 'cow-udder' (Fig. 2.15b). Even more surprisingly, South African wild bees construct honeycombs to this design (Fig. 2.15c).

It seems very probable that the "intestines" of both the Ndedema eland and the Lascaux bison are pictures of honeycombs and that what we see are the results of imposition of new art on old thus creating images, which can be variously interpreted

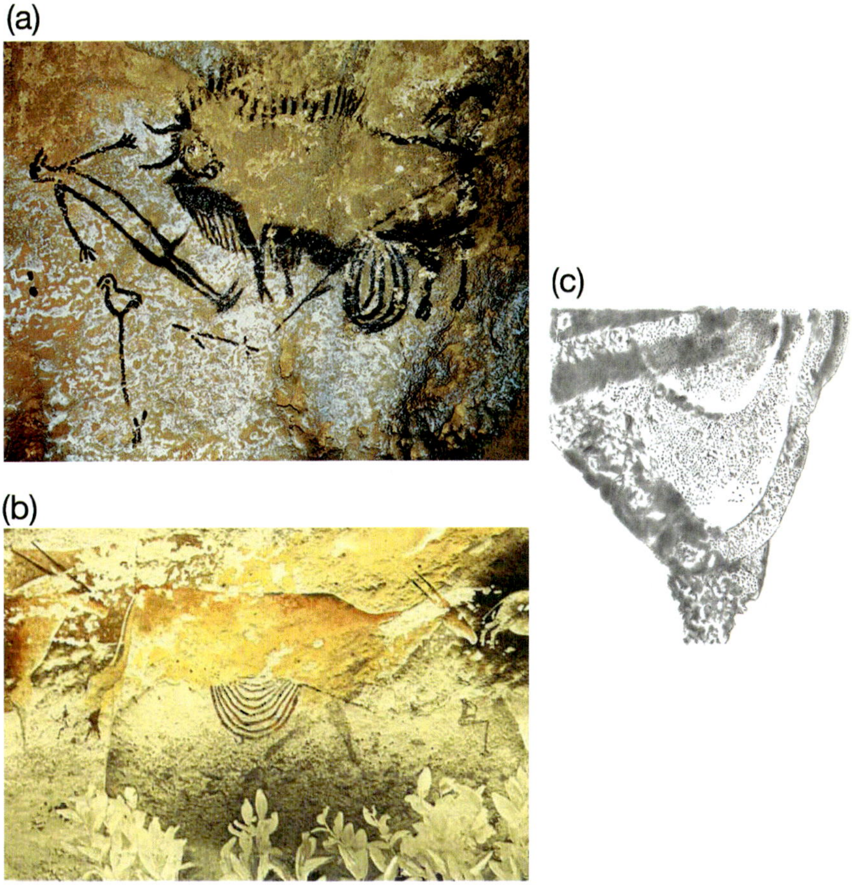

Fig. 2.15. **a** A gory scene from Lascaux. A bird-headed shaman is prone on the ground, his opponent's—the bisons—intestines are dangling below its belly **b** Less gory, but still a gory scene, from Botha's Shelter in Ndedema Valley **c** A palliative. A catenary bees' nest which disposes of all anguish (based on Pager, 1972)

and one of such interpretations is particularly attractive because it is dripping with gore.

The differences in size among similar looking elements, constitute texture gradients, which whether occurring naturally on the surface on which the picture is drawn, or created deliberately by the artist (brush strokes, portrayed details, etc.) affect perception.

Farmer McDonald sees the texture gradient formed by the rows of his cabbages. He notices that the cabbages, which are all similar in size and shape, appear to vary with distance: those more distant seem smaller. He would, however, be wrong to think it a rule, for the cue of apparent diminution is not absolute. More distant objects do not necessarily project smaller images into the eye than objects closer

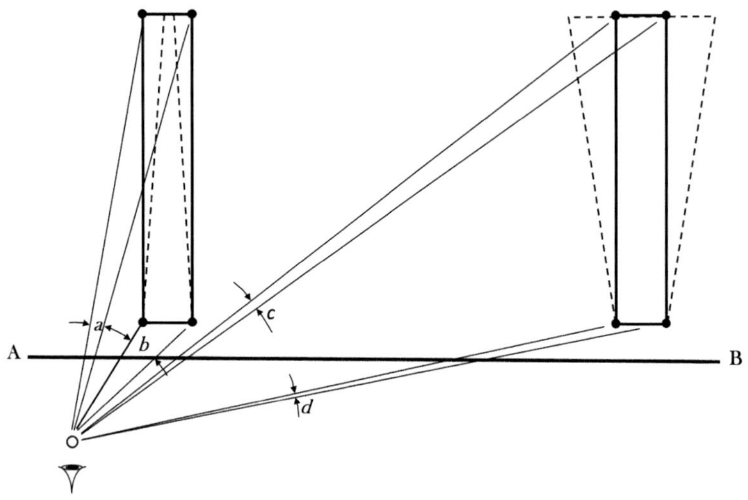

Fig. 2.16 Angles at the eye cast by two identical rectangles on the floor, one almost in front and the other laterally displaced

by. Artists in many cultures, but most notably Byzantine and oriental artists have acknowledged this, showing that such deviations from central convergence with depth found in pictures are not results of artistic fancy, but are evidence of visual experiences available to all.

The figure shows two identical rectangular arrangements of four markers placed on the floor in front of the observer (Fig. 2.16). Rectangle A is placed immediately in front of the 'eye'; rectangle B is placed well to the side. The angles which pairs of markers support at the eye and which therefore determine sizes of projections are shown by the dotted lines. In the case of rectangle A, the more distant pair of markers sustains a smaller angle ($a < b$)—an instance of *convergence* with increasing distance. In the case of rectangle B, however, the more distant pair of markers projects a larger angle ($c > d$) and therefore markers appear to *diverge* with increasing distance. This effect demonstrated on paper is readily observable in the laboratory[17].

Every surface on which the artist may wish to draw is made visible by its optical texture gradient. (The query as to how the artist could draw on an invisible surface is outwith the scope of the present considerations.) The texture gradient derives from the increasingly coarser appearance of a surface as the eye moves closer to it and increasingly finer appearance as it moves away. Thus a stationary eye looking at a flat surface which is physically uniform but is inclined away from the eye, such as farmer McDonald's cabbage field, sees a texture gradient of increasing fineness with distance. Such apparent change of fineness is independent of ease with which elements subject to the change can be identified. It is observable, and depictable, when they are paving slabs of a pavement or bricks of a wall, sand banks, or lawns.

The texture gradient seen by the observer depends on the physical texture gradient of the surface and the angle which the elements of the surface sustain at the eye. When the physical texture is uniform (isomorphic), as it is, for example, in the case of a flat brick wall, the observer looking normally at such a wall does not perceive the texture gradient as uniform but as changing symmetrically in all directions. This is the arrangement in which the view of the wall is the simplest possible—the informational load is least. The load increases when the wall is viewed at an angle and symmetry destroyed.

The perceptual system prefers to reduce perceptual load; to increase rather than to decrease symmetry, as Thouless's[18] experiment shows. In his experiment a subject sat in front of a screen. A projector placed above the subject's head casts a circular image on the screen. The subject, not surprisingly, saw it as circular. Rotation of the screen about its vertical axis through an arbitrary angle alters the shape of the projection to an ellipse with its horizontal axis longer than its vertical axis. The *apparent* shape seen by the subject is also an ellipse, and is a compromise between real shape that appears on the screen and the retinal shape which is not affected by the rotation of the screen and remains circular. There are only two vectors affecting the eye which determine the percept: The retinal image of the texture gradient of the screen and the retinal image of the projected shape. The 'true shape', measurable on the surface of the screen, is the shape obtained when the first of these is symmetrical.

The visitor to a picture gallery positions himself in front of a picture in a stance which maximises the informational redundancy of its surface's texture gradient, this is generally the stance in which the line of sight is perpendicular to the surface of the picture. Thus the picture's texture gradient suggests to the viewer whence the picture should be viewed[19].

There are pictures so conceived as to minimise the information their surface provides. They offer no guidance as to the viewer's stance. Perceived three-dimensionality of the depicted elements is unbridled. Their perceived shape varies with the viewer's stance. In such cases the artist may intervene and indicate the appropriate point of view. For illusionistic ceiling paintings found in some Baroque churches, this is done by marking a spot on the floor at which the viewer should stand; creators of panoramas confine viewers to viewing platforms. The method by which these two kinds of artist achieve the total effect differs. The artist cannot eliminate texture gradient of the surface on which he is painting, but he can decrease its visual presence to such an extent that it does not significantly influence the viewer's percepts. The illusionistic painter does so by placing the picture far from the viewer. The creator of panoramas uses an additional device of displaying his canvas on the inner surface of a cylinder, in the middle of which the viewing platform is erected. This ensures that they always remain at the same distance from the viewer's eye placed at the axis of the cylinder, and therefore show an unchanging texture gradient.

Andrea Pozzo (1642–1709) was the grand master of the illusionistic style. St. Ignazio's Church in Rome and the University Church as well as the Jesuit Church in Vienna contain outstanding examples of his work. The paintings on the walls of the church extend them upwards, creating perceptual space in which angels and putti hover. Seen from a designated spot on the floor of the church the depicted architectural elements seem real and are indistinguishable from the solid walls, columns and pilasters. If, however, the observer moves from the designated spot, his percept

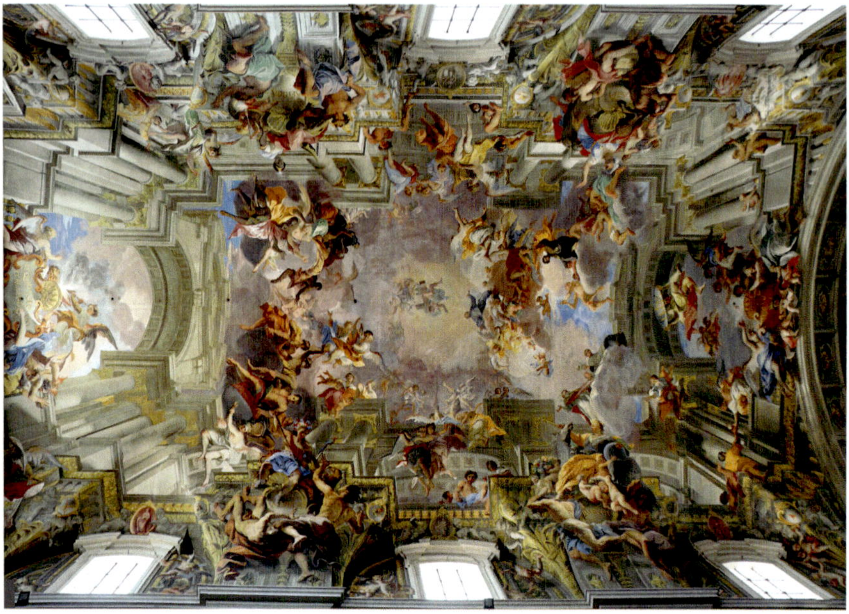

Fig. 2.17 Pozzo's Illusionistic Ceiling

changes, the painted architectural features, while still appearing real, distort and these distortions increase with the increase of the viewer's distance from the designated spot. Angels and putti do not seem as dramatically affected, but then an average viewer is unlikely to be familiar with angelic aerobatics and the shape of angelic garments (Fig. 2.17)[20].

The designers of panoramas, in an endeavour to banish such alarming experiences as those encountered by viewers of Pozzo's ceiling wandering from the designated spot, confine their viewers to a platform at the axis of a cylinder on the inner surface of which the picture is painted. The annular pace between the viewing platform and the cylindrical canvas contains objects which are extended into the pictorial space just as are the architectural features in Pozzo's paintings. Thus in "Panorama racławicka" in Wrocław, Poland, a real fence merges into a depicted fence. The viewer is not confined to one spot but is free to move on the central platform and to view, eye span by eye span, the entire picture. The viewer cannot evade the laws of optics entirely, the changes of stance do cause changes of the perceived shape, but these changes are not significant. The projected shape of a hut changes with stance, but so would projected shape of a real hut. A hut remains a hut. A tree too remains a tree and its change of shape is not even noticed. Trees are like Pozzo's angels greatly adaptable[21], (Fig. 2.18).

Ready acceptance of varying views, sometimes distorted views, of the hut in the Panorama is not evidence of great tolerance of the perceptual system, but of its drive to make sense of the environment. It is a drive which rules all normal perception.

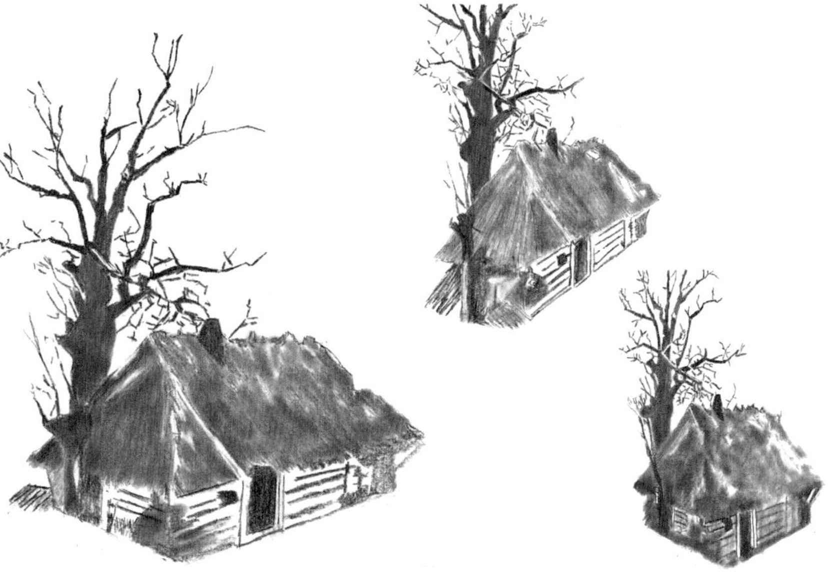

Fig. 2.18 Views of a hut, depicted on Panorama Racławicka, from two different stances, and, larger, a view from the intended stance. A hut remains a hut (based on photograph in Bartel, 1933)

Such variable percepts of objects are experienced daily and are accepted; a postman does not cease to be by turning around. The artist can reproduce all the poses. The artist enjoys this freedom and more, for artistic freedom offers not only opportunities to portray daily experiences, but also an opportunity to go beyond them by combining and compounding, creating the Pegasus, the Centaur, and the Gorgon—these, not flat on the canvas but in visual space. To achieve this ability to create perspectival paintings is needed.

The reviewed data show that a strong tendency to see depth in simple non-representational patterns, such as oblique lines prevails and that visual illusions attributed to perception of depth are seen by all (albeit at various intensities). Such illusory depth is most strikingly evoked by pictures drawn in perspective—to which we shall now turn.

Notes

1. Euclid, *Optika*. For translation see H.E. Burton (1945).
2. (Deręgowski and Parker, 1988), and cognate reports by these two authors 1992 and 1995.
3. Józef Chełmoński (1849–1914) a realist painter, is the author of both "Four in hand" and "Gossamer".
4. See (Wade, 2017), for "Early History of Illusions".
5. See Segall, M.H. Campbell, D.T. and Herskovits, M.J. (1966). Their's is the largest cross-cultural study of perception using illusion figures. It involved, in the case of the two illusions

discussed here, fifteen different adult samples and thirteen samples of children. The Horizontal-Vertical illusion was investigated with two kinds of stimuli, the authors having overlooked (Kunnapas, 1955) study which shows that T-form of the illusion figure is not the most suitable for investigation of their hypothesis.
The P.S.E. (Point of Subjective Equality) measure is the percentage of *discrepancy* at which half of the sample perceives the two compared elements to be equal. The means of this measure derived from responses of the groups in which both adults and children were tested were as follows: For the T-type figure: Adults: 17.8, children: 17.8. For the L-type figure: Adults: 9.0, children: 11.0. The discrepancy between two of the illusions experienced with two types of stimulus is that demonstrated by Kunnapas. It is noteworthy that children's P.S.E. in both conditions is not less than that of the adults, although their visual experience is lesser.

R.L. Gregory was particularly interested in the classical form of the Muller-Lyer illusion. He postulated that certain features are perceptually expanded when signaled as being further away, and that arrow-shaped corners do provide such signals. Representations of such corners are a feature of the illusion. Gregory illustrated his argument by pointing out the similarities of an inside corner of a room to the Muller-Lyer figure with inward pointing arrowheads and of the outside corner of a building to the Muller-Lyer figure with outward pointing arrowheads. (See Gregory, 1963; Gregory, 1997; Gregory, 2009). The question "To what extent daily visual experience of objects showing such features affects susceptibility to the illusion?" induced M.H. Segall, D.T. Campbell and M.J. Herskovits (1966) to conduct a cross-cultural investigation.

6. (Swift, 1733).
7. All adult groups reported by S. C. & H.
   Had P.S.E. greater than 0. Of children's groups only one (Suku from the Congo) showed no illusion susceptibility (P.S.E. = 0). Overall means of P.S.E. were: adults: 7.4, children: 9.7, which is contrary to expectations, since children's experience of the carpentered world is obviously lesser.

   The Muller-Lyer stimuli used in the experiment had mutually orthogonal fins, an acceptable, but borderline condition. Fins intersecting at an angle smaller than 90° do not relate directly to the experiences of orthogonal solids, such as cubes, cuboids and many artifacts (Deręgowski, 2013).
8. (Fry 1910).
9. (Mach 1897). See Fig. 1.3.
10. See Breuil and Obermeier (1935).
11. Comparison between the relative elevations of hooves of forelegs with hooves of the rear legs yields the following:

    i. Contorted animals (the forelegs are not congruent with the rear legs): 11 instances.
    ii. The two pairs of legs are in line and the animal lies on its flank: 2 instances.
    iii. The two pairs of legs are in line and are protruding upwards: 2 instances.
    iv. The two pairs of legs are in line and the animal is upright: no instances.

    The comparison excludes four 'curled' animals.

12 (Leason 1939), an artist, brought the problem of unnaturally arraigned feet of depicted animals, yet said to be engaged in such activities as trotting, bellowing, charging, etc. to general attention and pointed out that the problem of the feet as well as several other apparent lapses of draughtsmanship vanish when the depictions are seen as of cadavers, not of living animals. He supported his argument by pointing out that photographs of animal cadavers taken at slaughterhouses show configurations similar to those in Palaeolithic pictures. His is a convincing and lucidly made argument which seems to have been largely overlooked, perhaps because descriptions in terms of action such as trotting, bellowing, charging, etc. are more attractive than a simple statement that one is seeing a corpse. The present evaluation is made using Breuil (1935) book. It rests on those depictions of bisons which could be said to show upright animals on clearly visible legs. J.B. (Deręgowski 2017) presents a brief report concerning the extent of spatial concord between overlap of the legs and size of hooves in Altamira paintings

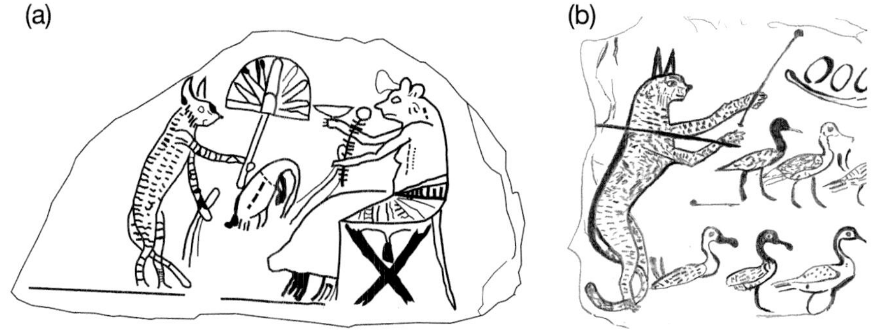

Fig. 2.N1 **a** An ostracon; an Egyptian cat, a revered animal, pays a humble homage to a mouse. After Swindler ('1931) **b** Another (?) cat, husbanding geese. (After Michałowski, 1969)

and finds such concord to be present. Not only does the devil lodge in the detail, but his very lodgings are informative.

13. Schafer (2002) discusses this issue at some length, pressing the point made in the citation. His admirable and scholarly work is therefore probably largely responsible for the prevalent acceptance of this, mistaken view.
14. (Deręgowski 2017).
15. The spokesman for the Egyptian tradition, an obedient follower of the received art code, omits to mention ostraca, potsherds and flakes of limestone, on which some artists felt free to express themselves and did so with gusto, as the picture of a cat paying homage and bearing humble offerings of fish to a grandly seated mouse (Fig. 2.N1a. after Swindler, 1931) shows.

    The cat's strange pursuit forms a puzzling link to an ostracon held at the Cairo Museum (Fig. 2.N1b, based on Michałowski, 1969, on which a cat, in a similar pose, is husbanding geese. The obvious questions arise,are the two cats related? Is it the same cat? These questions fall outwith the scope of this wee book.
16. See (Pager 1972), who in his *Ndedema* juxtaposed a photograph of a catenary bees' nest with that of a painting of an eland whose "intestines" are spilling out. He postulated (Pager 1974) that overlapping found in Bushman pictures had a magico-religious origin, and that bees (and allied subjects) were frequently involved in overlapping because of their "great esoteric significance." See (Crane 1983) for a broad discussion of beekeeping in various cultures.
17. (Guczalski 2012) in his extensive examination of the concept of perspective dismisses inverse perspective and hierarchical perspective (whereby objects, generally persons, of greater importance, are depicted as larger than objects of lesser standing) as misnomers. This decision is questionable. The origins of these perspectives are clearly distinct,the former derives from perceptual experience, the latter from social convention. We are concerned solely with the former. The dismissal seems overhasty as far as inverse perspective is concerned because, just like the convergent perspective of Leonardo da Vinci's pane, it arises in the course of normal operation of the visual system (Ten Doesschate 1964; Deręgowski and Parker 1992; Deręgowski et al. 1994). It occurs when stimuli viewed are laterally displaced. In those realms it replaces convergent perspective. There is inevitably a range when parallel perspective is seen. Such perspective is used in some oriental art. In the twelfth century magnificent Japanese scrolls documenting a fire at a temple it is the only form of perspective used (Fig. 2.N2). Both 'non-convergent' forms of perspective have been used in various artistic traditions. Such depictions are, however, seen as odd when looked at centrally by people used to 'convergent' perspective, especially so when the depicted objects are thought to have rectangular shape.
    Convergent perspective looks odd to people from cultures where it is not commonly used. Yoshio (Markino 1912) reports that his father, a cultured Japanese gentleman, thought drawing

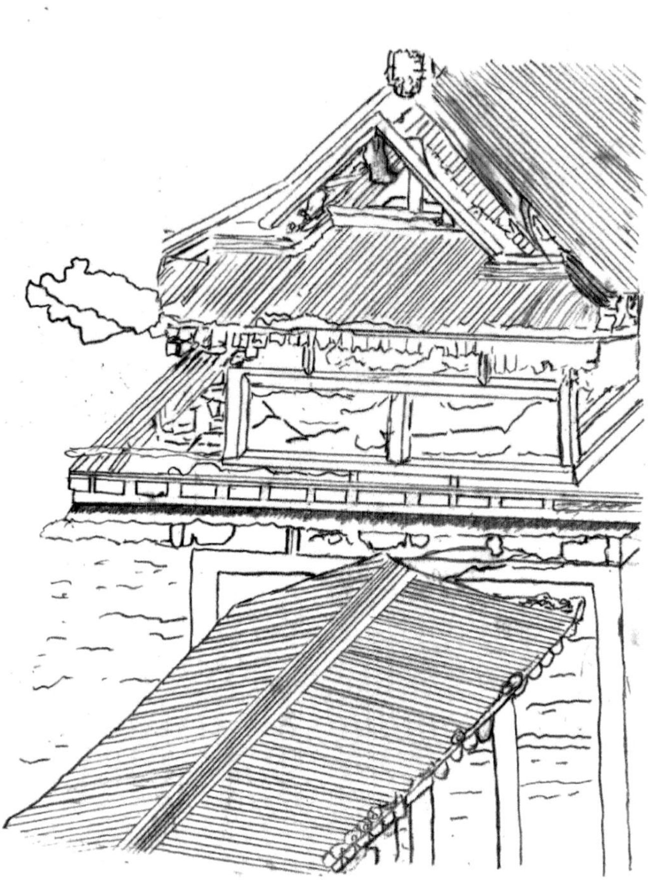

Fig. 2.N2 Element from a Japanese scroll painting of a temple. Parallel perspective rules!

in perspective of a box to be crooked. Nine years after after much exposure to European art (he lived in London), he thought the same picture to be "perfectly right" (see Gombrich 1962).

Perspectival cues are generally enshrined in elements of supposedly known shape, such as floor rugs and table tops, or elements known to be equal, such as the telephone poles or in architectural features. It is these that a zealous academic perspective seeker uses to determine whether the picture is drawn in convergent perspective and having found convergence declares it to be so. The claim is questionable for as Bartel's experiments show different convergent depictions of a line of telephone posts evoke different percepts. When the shape of the depicted object is unknown such reservations do not apply. The king (Fig. 2.N3) would be as uncomfortable in his bed drawn in 'convergent' perspective as he is in his 'inverse perspective' bed of the thirteenth century Aberdeen Bestiary.

18. Thouless (1972).
19. In his foreword to (Pirenne 1970) book Polanyi provides an illustration of the function of subsidiary awareness. It is like a pointing finger of which we are not unaware but which guides us how to perceive the depiction. A photograph viewed from a different stance than the point from which it was taken is not seen as distorted (if this happened passport photographs

(a) (b)

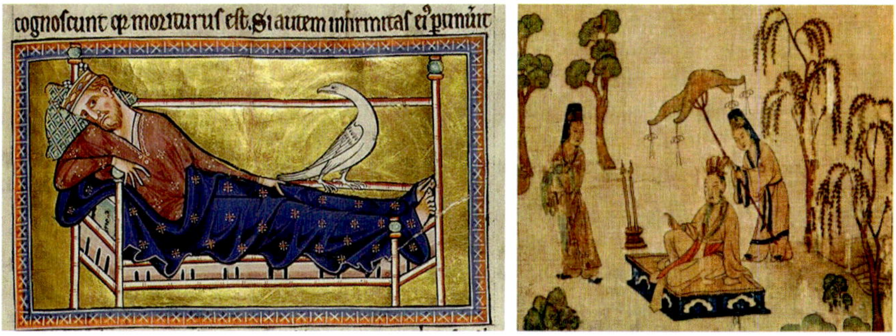

Fig. 2.N3 A king in the royal bed, an instance of divergent perspective. Aberdeen Bestiary 13th cent., and an earlier painting from a different culture by Gu Kai Zhi (Fig. 4.7) from the end of the fourth century showing an enclosure, one suspects to be rectangular, in reverse perspective. (See Chap. 4 for further comments on Gu Kai Zhi's picture.)

would not be very helpful), as the finger of subsidiary awareness embodied in its surface guides the viewer. This guidance does not extend to photographs of photographs whose surfaces are not seen, therefore objects which they depict look grossly distorted. (Pirenne presents two interesting illustrations of this effect). Not only the surface of the pictures affects the percept, but so does, Pirenne says, the boundary of the picture, its frame. The boundaries, as their role in forming typical contours shows, influence formation of percepts greatly. The painter not only paints a picture, he also decides on the shape of its frame.

20. For exploration of the perceptual effects of illusionistic paintings, see (Pirenne 1970).
21. For exploration of panoramic art, see (Bartel 1958).

References

Bartel K (1958) Perspektywa Malarska. Państwowe Wydawnictwo Naukowe, Warszawa

Breuil, H., Obermeier, H.: The cave of Altamira at Santilana del Masr. Topografia de Archivos, Madrid, Spain (1935)

Crane, E.: The Archaeology of Beekeeping. Duckworth, London (1983)

Deręgowski, J.B.: On the Muller-Lyer illusion in the carpentered world. Perception **42**, 790–792 (2013)

Deręgowski, J.B.: Perspective on perspective in palaeolithic art. Rock Art Research **34**, 106–107 (2017)

Deręgowski, J.B., Parker, D.M.: On the changing perspective illusion within Vermeer's "The Music Lesson." Perception **17**, 13–21 (1988)

Deręgowski, J.B., Parker, D.M.: Three-space inference from two-space stimulation. Percept. Psychophys. **51**, 397–403 (1992)

Deręgowski, J.B., Parker, D.M., Massironi,: Perception of spatial structure with oblique viewing: an explanation of Byzantine perspective. Perception **23**(5), 13 (1994)

Elmslie, W.A.: Among the wild Nguni. Oliphant, Anderson & Ferrier, London (1901)

Euclid Optika trans. Burton, H.E.: Journal of the Optical society of America 35: 357–372. Also: http://philomatica.org/wp-content/uploads/2013/01/Optics-of-Euclid.pdf, And (in Greek) http://www.wilbourhall.org/pdfs/Euclid vol VII.pdf (1945)

Fry, R.: Bushman paintings. The Burlington Magazine for Connoisseurs **16**, 334–338 (1910)

Gombrich, F.H.: Art and Illusion, a study in the psychology of pictorial representation. Phaidon Press, London (1962)
Gregory, R.L.: Distortion of visual space as inappropriate constancy scaling. Nature (London) **119**, 678 (1963)
Gregory, R.L.: Eye and Brain. Oxford University Press, Oxford (1997)
Gregory, R.L.: Seeing through illusions. Oxford University Press, Oxford (2009)
Guczalski M (2012) Perspektywa. Forma symboliczna czy naturalna? Homini, Kraków
Kunnapas, T.M.: An analysis of the vertical-horizontal illusion. J. Exp. Psychol. **49**, 134–140 (1955)
Leason PA (1939) A new view of the Western European Group of quaternary cave art. Proceedings of the Prehistoric Society for 1939 New series vol. V:51–60
Mach, E.: Contributions to the analysis of the sensations. Open Court Publishing Company, Chicago (1897)
Markino, Y.: When I was a child. Constable, London (1912)
Marmor, M.F., Ravin, J.G.: Artist's eyes: Vision and the History of Art. Abrams, New York (2009)
Michałowski, K.: The Art of Ancient Egypt. Thames & Hudson, London (1969)
Pager, H.: Ndedema. International Scholarly Book Services Inc., Portland, Oregon (1972)
Pager, H.: Rock paintings in South Africa showing bees and honey hunting. Bee World **54**, 61–68 (1974)
Pirenne, N.H.: Optics, painting and photography. Cambridge University Press, Cambridge (1970)
Schafer, H.: Principles of Egyptian Art. Griffith Institute, Oxford (2002)
Segall, M.H., Campbell, D.T., Herskovits, M.J.: The influence of culture on visual perception. The Bobbs-Merrill Co, Indianapolis (1966)
Swift J (1733) Poetical works. Davis HJ (ed.) (1967) lOxford University Press, London
Swindler, M.H.: Ancient Painting. Yale University Press, New Haven (1931)
Ten Doesschate G (1964) Perspective, Fundamentals, Controversials, History. B De Graaf, Nieukoop
Thouless, R.H.: Perceptual constancy or perceptual compromise. Aust. J. Psychol. **24**, 133–140 (1972)
Tongue, H.M.: Bushman paintings. Clarendon Press, Oxford (1909)
UNESCO: Simple reading material for adults: Its preparation and use. UNESCO, Paris (1963)
Wade, N.: Early history of illusions. In: Shapiro A, Todorovic D (eds.) Oxford Compendium of Visual Illusions. Oxford University Press, Oxford (2017)

Chapter 3
On Perspective in Many Guises

The notion that a perfect picture presents the viewer with stimulation identical with that presented by the model, is at the core of Leonardo da Vinci's[1] definition of perspective:

> Perspective is nothing else than seeing a place (or objects) behind a plane of glass quite transparent, on the surface of which the objects behind that glass are to be drawn. These can be traced in pyramids to the point in the eye, and these pyramids are intersected on the glass plane.

There are two ways in which the adjective "quite" of the above definition may be understood. It may mean, (a) "to the utmost or most absolute extent", but it may also mean (b) "to a fairly significant extent". A dictionary provides the following examples of the two uses: (a) "this is quite out of the question", and (b) "it's quite warm outside". [2] In the context of Leonardo's definition of perspective, the adjective may indicate either that the glass is perfectly transparent, and therefore it is not visible, or that its impaired transparency ensures its visibility and the percept incorporates both the projected image and the image of the pane's surface. Painting on Leonardo's perfectly transparent pane would create a perfect picture such that, on its removal from the easel the eye, provided it had remained in one fixed position, would not perceive any change. If such a perfect picture were presented to the eye in any other position or orientation than that in which it was painted, the percept would no longer be that of the model. This does not mean that it would become meaningless. As shown in the previous chapter, portrayals on Pozzo's Baroque ceilings and panorama walls (both of which present no texture gradients, whether seen from the designated stance or some other stance in its neighbourhood) do not turn into a meaningless melange on the eye's removal from the designated stance. They merely change the shape (and therefore, sometimes, the nature of the portrayals); buildings may be enlarged or shrunk, trees may change shape and so may angels.

Most traditional pictures are such that in terms of Leonardo's definition they are painted on the glass which is "transparent to a fairly significant extent" because their surfaces are visible, they have texture gradients. Such pictures provide two distinct kinds of stimulation, that created by the artist and that which was present

J. B. Deręgowski, *On Picture Making and Picture Seeing*,
Vision, Illusion and Perception 4, https://doi.org/10.1007/978-3-031-23348-7_3

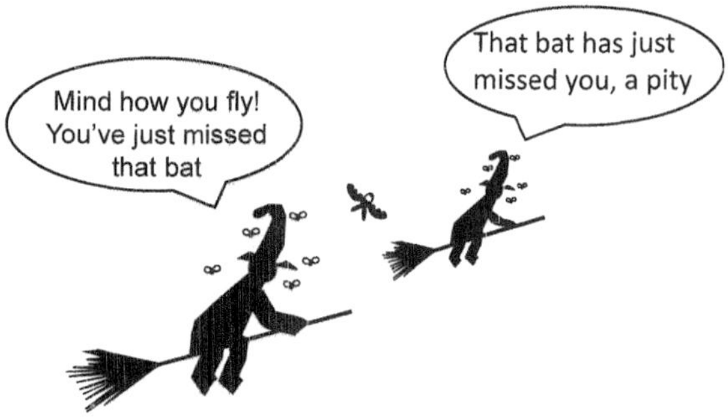

Fig. 3.1 Two witches on their way. "Mother and daughter, perhaps, certainly not monozygotic twin sisters", says Upright Geometer

even before the work of art was begun. This distinction is, however, explicit only in simple pictures; there is the surface (the rock face, say) on which the artist depicts and there are lines of pigment or engraved grooves forming the image. Once the artist incorporates elements of the surface in his schema, as he may do by using a natural bulge of rock face to accentuate a part of an animal's depiction, or by forming the paint on the canvas so as to alter the natural texture gradient of the surface, the two elements—the surface and the portrayal are seen as a single stimulus.

Perception, other than perception by specialists trained to see the world differently than people in the street, does not entertain such a distinction. The eye receives stimulation and tries to make sense of it; both to classify the depiction and to perceive pictorial space. This is the very purpose of the eye's existence.[3]

Such pictures are seen differently by the Ardent Illusionist who takes cognizance of all the depth cues presented by the scene behind the pane and dismisses entirely the realistic cues of the flatness of its surface, and an Upright Geometer who sees the picture as flat and therefore all the markings as co-planar so that, for example, the size of depiction is not a cue to pictorial depth and the two witches on their broomsticks in Fig. 3.1 are perhaps mother and daughter but certainly not twin sisters.

Much can be gained from the cooperation of the Ardent Illusionist with the Upright Geometer. If the Illusionist points out to the Geometer the line representing the horizon and some other lines which, albeit oblique on the picture's surface, are projections of lines normal to the horizon, much, as Bartel[4] has demonstrated, can be gained. Bartel's studies consist of a number of experiments in which he used a geometrical method of his devising for determination of the relative magnitudes of depicted distances. He refers to this process as *restitution.*

Restitution takes account of the obvious, yet oft overlooked issue of consistency of pictorial depth cues; the rate of convergence of the rails of a track extending away from the viewer should be geometrically consistent with the rate of compression of spaces between the sleepers (ties). The artist may acknowledge that objects further away are seen as smaller and depict them thus, yet fail to size and space them correctly in the picture. A depicted colonnade may show columns of varied heights at spans incompatible with the rules of geometric perspective. A perspectivally truthful depiction should take account of the relationship between two phenomena: 1. Convergence with distance – as Farmer McDonald's rows of cabbages seem to converge,

2. Compression of space—those of Farmer McDonald's cabbages which are further away seem smaller, as do the spaces between them. Errors in compression are more easily noticeable when elements depicted are supposed to be identical (such as, say, telephone posts), but less easily when they are different (say, leafy bushes).

Errors of convergence are explored in every popular text on perspective. They are easily apprehensible and lend themselves to immediate convincing demonstration by drawing lines on pictures. They have dominated thinking about perspective to the extent that it is sometimes implied that mere presence of convergence in a picture indicates that the artist had perfect grasp of perspective, although geometrical analysis of the picture may reveal that he did not, but that he depicted convergence-with-distance to evoke perception of depth and paid scant attention to compression of space. Masolino's da Panicale's fresco, "Presentation of St. John's Head" at The Baptistry, Castigilone d'Ola (Fig. 3.2) depicts a long row of columns extending away from the viewer. Columns converge towards the horizon and those, apparently further away, are systematically drawn smaller. This is the dominant stimulus for perception of pictorial depth. It works in spite of the fact that the picture does not fully abide by the rules of perspective; it does not show columns as evenly spaced, as one would expect them to be, but as increasingly further apart; the most distant span is nearly eight times the size of the nearest. Such expansion of pictorial space with apparent distance from the viewer is widespread, it is also found in the paintings of Pieter Jansz Saenredam (1597–1665), whose portrayals of church interiors are held to be unsurpassed. This predominance suggests a perceptual discrepancy between the geometrically consistent elements of perspective.

The two vectors affecting the perception of Farmer McDonald's cabbages emerge in Marsolino's fresco: rows of cabbages appear to converge with increasing distance and spans between them, kept constant when planting, as well as the cabbages themselves, appear to shrink. The depicted relationship between these changes is not that presented on Leonardo's pane. Had Masolino, unaided by the glass screen, painted a field of cabbages, the cabbages in rows in his picture would not decline in size, in harmony with the rate at which rows of cabbages would converge; although both effects would evoke pictorial depth, they would evoke it at different intensity. Such discrepancies are to be found, even in depictions of chess boards; presenting, as they do, square arrangements of squares. The artist may let this dominance of squares remain unchallenged, and does indeed often do so. The chess board is then depicted with the mutual orthogonality of its square boundaries unquestioned and appears to be in the vertical plane, as do many centrally placed rugs in oriental miniatures (see Fig. 1.15). Such portrayals of chess boards are commonplace, but when the artist endeavours to show a chess board in depth he needs to attend to both the apparent convergence of parallel lines extending away from the eye and the apparent compression of squares between these lines. (In a picture of a railway track this would entail convergence of the rails and compression of the inter-sleeper (inter-tie) spaces.) Difficulties facing the artist are shown by the two pictures in Fig. 3.3. In the earlier, by Albertus Pictor, at a church near Stockholm, Death is obliged to play a game of chess on an unconventional chess board, in the later, an illustration of a poem about chess, shows difficulties experienced by another artist. The unexpected expansion

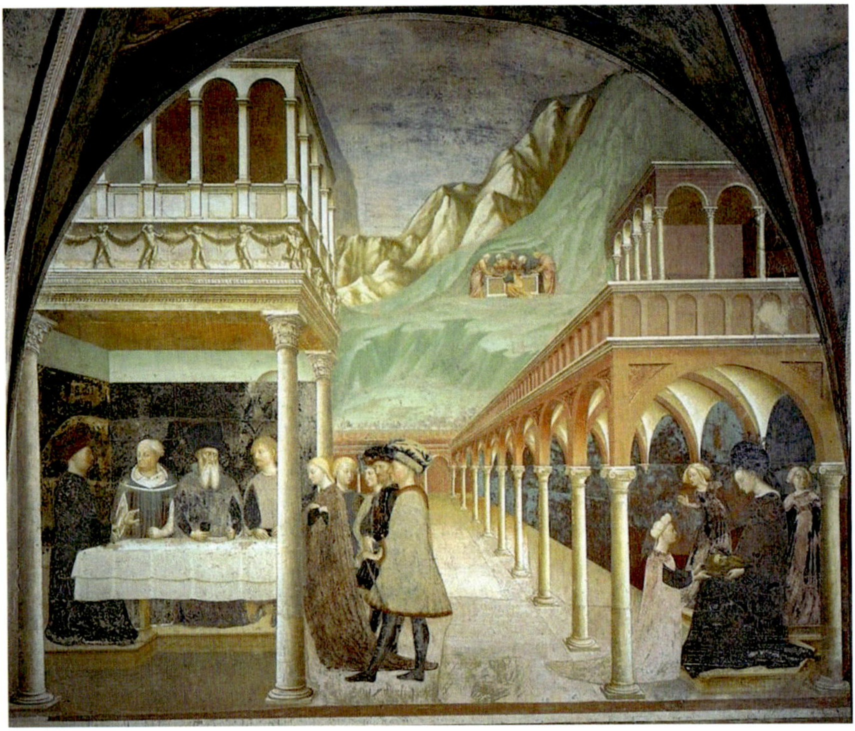

Fig. 3.2 A Masolino da Poincale's fresco (courtesy this image was downloaded from https://commons.wikimedia.org/wiki/File:Masolino_-_Banquet_of_Herod_-_WGA14245.jpg and is in the public domain, free of copyright (see information on that page)). Pictorial depth cues presented by the colonnade are, in true perspective terms, incompatible. The rate of convergence of the line connecting the columns' capitals with the line connecting the columns' plinths implies a greater rate of spatial compression than the rate implied by depicted spans of, presumably, evenly spaced columns. Bartelian restitution shows that the most distant span is about eight times the size of the nearest span (see Bartel, 1958, Fig. 236). This discord is apparent to some viewers who say that when monocularly concentrating on the row of white columns they do not see them as all co-planar, that the most distant columns deviate from the plane of the less distant. (When *making sense* of a picture we tend to override such local inconsistencies.) If the reader were to try to restitute the layout of the floor tiles in Vermeer's *The Music Lesson,* he would find no analogous discrepancies

of the more distant rows of the chess board is the artist's acknowledgement of the constancy principle described in the introductory paragraph of the second chapter. Leonardo's geometrical model of perspective does not take account of this psychological effect. The relationship between these two vectors: the rate of convergence and the rate of compression, is the foundation of Bartel's method, considered below.

Imagine a barren plain stretching in front of the artist; featureless, it extends towards the horizon, to which a single straight road leads. A line of telephone poles supporting two wires accompanies the road.

Our artist sets up his Leonardo's pane in the middle of the road and looks through it, with one eye from a fixed position towards the horizon. The horizon is a single horizontal line and all the linear features that are normal to the plane of Leonardo's

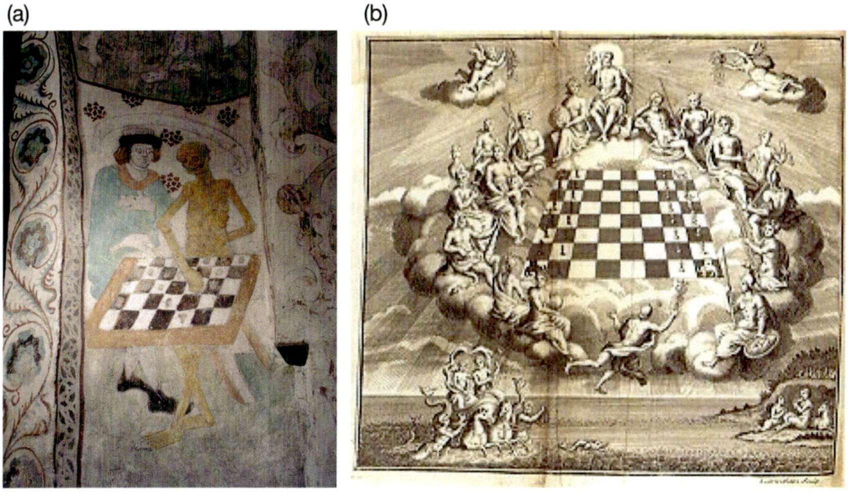

Fig. 3.3 **a** Painting by Albertus Pictor at Taby church near Stockholm between 1480 and 1490. **b** An illustration to Marcus Hieronymus Vida's poem on the game of chess *Scacchia Ludus*, published in 1527

pane, the edges of the road, the telephone wires, appear to converge to a single point on the horizon in front of the artist's eye; the telephone poles appear to grow shorter with distance, as do the distances at which they are spaced, all in accordance with the rules of perspective, and in accordance with these rules both the poles and the spaces between them vanish at the point in front of the eye.

Our artist traces in the road and the telephone line. Of the telephone posts he sketches in but two, one close by and another at a fair distance, knowing that since the intermediate poles are, in accordance with the Post Office rules spaced evenly and all the poles are of the same size, he, familiar with the rules of perspective, will be able to draw them in at leisure. The artist takes home a drawing (Fig. 3.4), which, fortunately, is also the stimulus used in many of Bartel's experiments.

Bartel's restitution of pictures having a clearly defined horizon and a set of vertical elements, such as telephone posts or columns, enables one to derive from portrayals true *relative* span sizes. Comparison of those with those known to prevail provides a measure of the extent to which the artist has abided by the rules of geometrical perspective. The spacing of the telephone poles in his complete drawing, restituted using Bartel's procedure, will show whether he is indeed an expert draughtsman of perspective.

The method of restitution can be used to derive from the distances in the picture plane among *depictions* of items, the *relative* distances among them in the real space, and to compare them with spacing of depictions of elements whose relative spacing in the real space (or whose putative spacing) is known.[5] Suppose that a landscape shows a road with parallel line of telephone posts stretching away from the viewer. If the picture is drawn in perspective one can ask: How do relative distances between

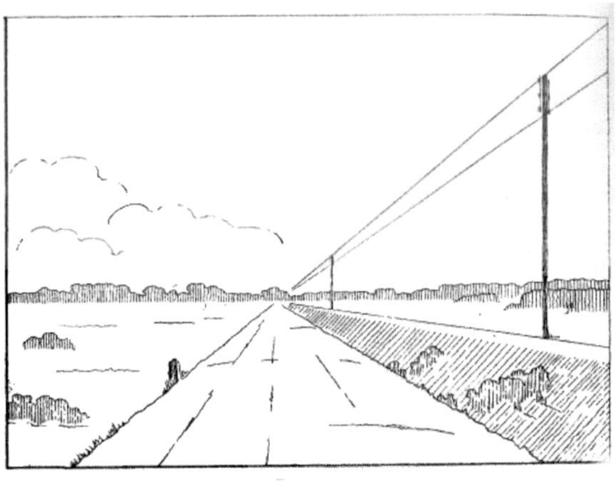

Fig. 3.4 A bleak view. Lines converging towards a point on the horizon evoke perception of depth. Perception of depth can be also evoked by compression of space shown e.g. by changing spans between telephone posts. In perspective of Leonardo's glass these effects are strictly correlated. This correlation is not found in many pictures. The relationship between these vectors is central to Bartel's investigations. (From Bartel, 1958)

depicted posts compare with the relative real distances? Or, if one wishes to depict another line of telephone posts, to ask: At what distances should the new posts be drawn, given the statutory distances between the real posts?

Bartel's studies

All Bartel's studies were carried out at the Technical University of Lwów (Poland) where Bartel held the chair of Descriptive Geometry. The subjects used were students. There were two distinct studies; the first of perception of objects in space and their depiction, and the second of perception of pictures.

The first study

Three of the experiments of this study were carried out in a specially constructed laboratory (Fig. 3.5a). Subjects were seated at a small table and in the first and the third experiment drew on a response sheet showing a perspective view of the arrangement of the walls in the room (Fig. 3.5b). They drew an array of black staves hanging on the white wall to their right. In the second experiment subjects instructed a laboratory assistant how to place the staves on the wall in order to create an evenly spaced array.

The array presented in the first experiment consisted of six staves defining five equal spans. The array used in the third experiment had spans equal to the means of spans recorded in the second experiment, which were judged by subjects to be equal.

Figure 3.6 summarises the results obtained in the three experiments. The results of the first and the third experiment are presented in the restituted form; they show spatial spans which would obtain given perspectival drawings, and not those in the

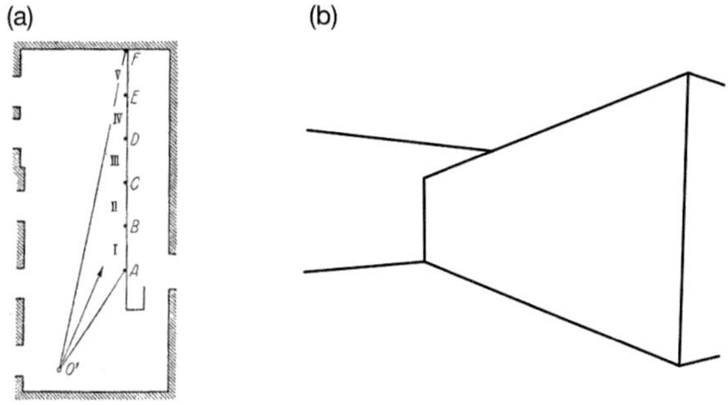

Fig. 3.5 **a** Plan of the laboratory with the built-in display wall on which staves were hung. The table at which the subjects sat was at O; A, B, …F are the staves shown here as evenly spaced. Consecutive spans, counting from the observer, are denoted by Roman numerals. Stave F was in the corner formed by the display wall and the back wall of the laboratory (Bartel, 1958, Fig. 185). **b** Response sheet used for the first and the third experiment showing the laboratory as seen from point O. The trapezium represents, in perspective, the display wall, top and bottom edges of display wall converge to a point on the horizon. The most distant stave is at F. (based on Bartel, Fig. 186)

plane of the drawings (see Appendix 5 for the procedures used). The results of the second experiment are the means of spans set on the wall.

All three measures show steady increase of spans with the distance from the observer. The drawings show that students intended to represent space. The discrepancy of results between the second and the third experiment reveals visual incompatibility of locations of objects in real and pictorial spaces. In the second experiment staves are set in the real space of the laboratory wall, in the third experiment their depictions are placed on a piece of paper showing perspectival view of the wall. Bartel describes the effect as follows.

> *Transformations of percepts (A1) (resulting from observation of real objects) into drawings, and re-transformations of drawings into percepts (A2) are not in general reversible and do not result in precise identity A1 with A2.*

Let the artist beware! Let the psychologist using pictorial tests beware!

Bartelian restitution can also be used to compare true spans with the spans derived from the drawings. In the first experiment this involves comparison of standard equal spans with spans obtained by restitution. Application of this technique to the data showed that the true spans beyond the second span were consistently smaller than the spans implied by the drawings. Analogous results were obtained when the mean true spans of the second experiment were compared with the spans of the third.

The fourth experiment of the study was concerned solely with pictorial space; students were asked to insert, in Fig. 3.4, between the two posts shown, four posts, bearing in mind that telephone posts are evenly spaced. It was found that the spans further from the viewer were made relatively larger – a result in agreement with the findings just discussed.

Bartel's second study

This study is concerned with perception of depictions.

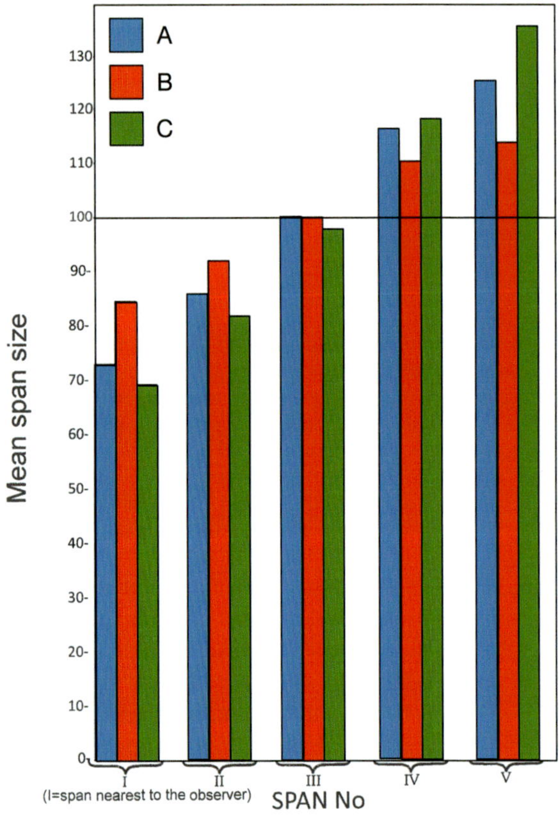

Fig. 3.6 Restituted spans and set spans. Line n is drawn on assumption that the total length of five equal spans is 500 units. Columns A, B, and C, show how this total length is partitioned by the subjects' restituted responses under different conditions: (A) First experiment, (B) Second experiment, (C) Third experiment. Data derived from Bartel (1958), Figs. 157, 189 and 190

Students were reminded that telephone posts were evenly spaced and were then invited to judge correctness of perspective in three pictures. The pictures differed only in the spacing of telephone posts (Fig. 3.7). Posts of picture A are spaced in accordance with the rules of geometrical perspective, just as they would appear on Leonardo's pane; spacing of those in picture B was that indicated by the drawings made in the last experiment of the previous study; spacing of the posts in figure C was at the means of the spacings in figures A and B, these means approximate closely to the means obtained by Bartel in the first experiment. The picture showing Da Vinci's geometrical perspective (Fig. 3.7a) was thought to be correct by four out of 81 students, the picture showing the poles spaced as indicated by the results of the fourth experiment of the first study (Fig. 3.7b), by twenty students, and the picture showing the poles at distances intermediate between the above two (Fig. 3.7c) by fifty-seven students.

Bartel's observations are consistent. The spans between evenly spaced staves are seen as increasing with distance from the observer, and a picture which shows telephone posts so placed that, on restitution its spans would yield an incremental series is judged to be the correct portrayal. Perceptual perspective embodies an incremental

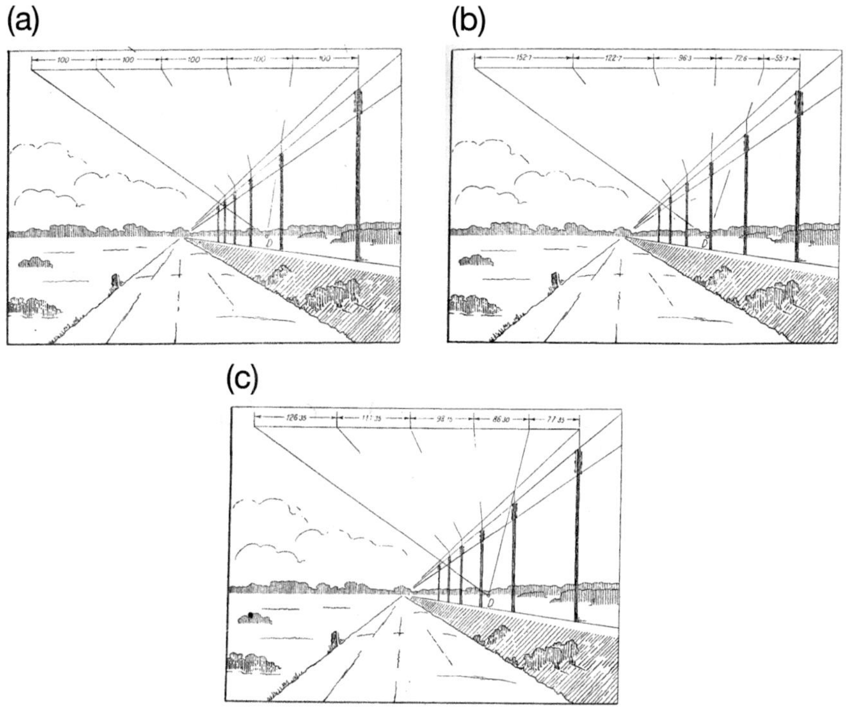

Fig. 3.7 Three depictions of a road with evenly spaced telephone poles: **a** In true geometric perspective. **b** As drawn by students. **c** With spans of mean lengths derived from spans of a and b

rule not recognised by geometrical perspective of Leonardo's pane. Geometrical perspective offers an approximation to the manner of seeing. This approximation does not take account of the incremental effect (Bartel's Law), which empirical evidence reveals, but it is robust and offers guidance to the artist wishing to create an 'as seen' picture. Notwithstanding this merit, it has provoked very verbose condemnations by those whose judgements are not based on empirical data and guided by reason.[6]

The relationship between the spatial constancy effect (which is present in both true and pictorial space) and the subsidiary awareness (which is present only in pictorial space) is obscure. One would expect subsidiary awareness to moderate the spatial constancy and therefore the results of the replication of Bartel's experiment[7] in which the staves were free standing to confirm this when compared with the results of Bartel's second experiment in which the wall was an obvious source of secondary awareness. However, other differences in experimental setting preclude such comparison.

The discord between geometry and psychology[8] documented by Bartel's experiments which occurs within the zone of central vision, has not been generally commented upon, unlike the discord occurring when a broader view of Leonardo

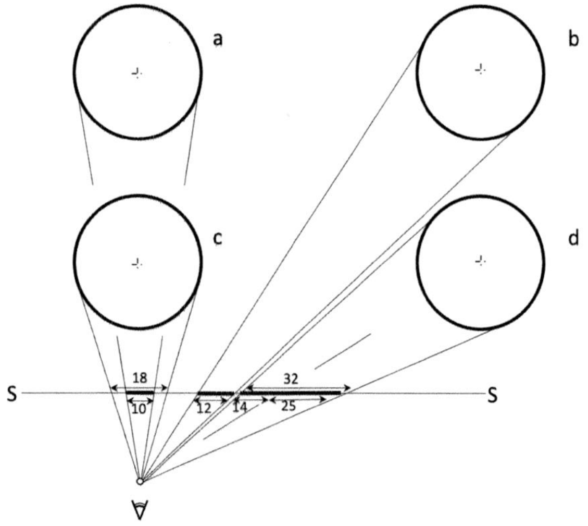

Fig. 3.8 Plan view of four cylindrical columns and a vertical screen (S). Columns *a* and *d* are at the same distance from the eye (O) but projection of column *d* is about three times the width of that of column *a*. The difference between projections of columns *d* and *c* shows the effect of lateral displacement, notwithstanding the change of the distance from the eye

da Vinci's pane is taken. The figure (Fig. 3.8) shows a bird's eye view of four identical vertical round columns a, b, c, and d placed in pairs in lines parallel to that of the pane in front of the eye. Columns b and d are at the same lateral displacement from the line of sight. Projections of all the columns onto the screen S–S are shown. (The effect is clearly related to that explored by the experiment involving rectangular arrangements of lights described in the second chapter.) Leonardo da Vinci was aware of the problem. His advice to painters is succinct: "It is well, therefore, to avoid such complex perspective and hold to simple perspective which… as much as possible portrays objects in their proper form".[9] He himself abided by this rule. He knew that a sphere, unless placed immediately in front of the eye, projects as ellipse—he is said to have never depicted a sphere.

Such considerations, it seems, guided Raphael when painting frescos in the Vatican, for he depicted two laterally displaced spheres by circles, not by the ellipses, that would have represented them on Da Vinci's pane. The fresco "School of Athens" is reproduced and commented on in, probably, every treatise on perspective. The most noteworthy and pioneering commentary is that of J. de la Gourniere, who not satisfied with merely stating the obvious, investigated the phenomenon by replacing circular portrayals of the spheres on a reproduction of the picture with correctly aligned ellipses. The modification, he found, did not improve the picture, on the contrary, it made it appear less convincing. De la Gourniere advised those who doubt this finding, to replicate the experiment. He was convinced that spheres should be represented by circles: "… *nous tenons pour certain qu'on dont representer une sphere par un cercle*".[10] The unique typical contour of the sphere dominates.

Figure 3.9 elucidates the 'distorting' effect of Leonardo Da Vinci's pane which de la Gourniere investigated. Three cubes, X, Y, and Z, are viewed through the pane. The cubes are aligned parallel to the pane with their centres co-planar with the eye. They differ in their lateral displacement. X is in front of the eye, Y is removed to

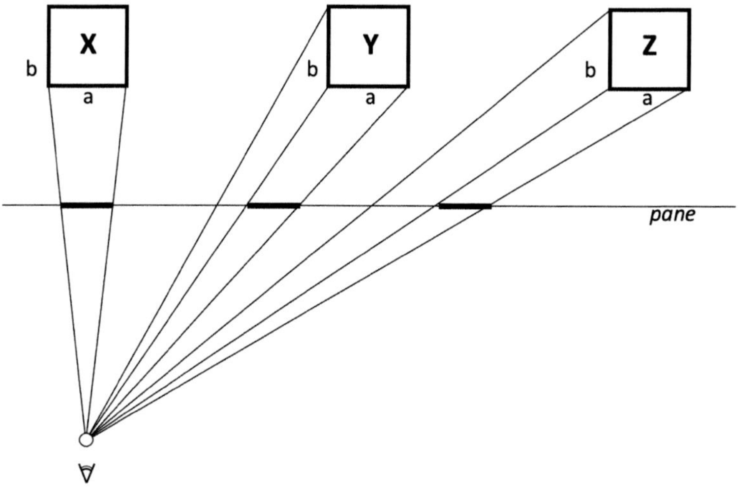

Fig. 3.9 A view from above of three identical cubes and Leonardo da Vinci's pane. Projections of faces *a* are of constant width, but those of faces *b* increase with lateral displacement. Change of the total widths of the projections is therefore due to changes of projections of faces *b,* if the cubes were replaced by cylinders or spheres, tangents, from the eye to the models, would change with lateral displacement (becoming more nearly parallel) the chords defined by them would become longer and the projected widths of the models would increase with displacement

the right, and Z even further to the right. Projections of faces 'a' on the pane are not affected by lateral displacement, they are all of the same size. This is not so in the case of faces 'b'. Face b of cube X is not projected on the pane at all, faces 'b' of cubes Y and Z are, and projection of Z is larger than that of Y. Thus, while projections of faces 'a' parallel to the pane are not affected by lateral displacement, projections of faces 'b' increase with displacement.

The gist of the observation is that a spatial component parallel to the Pane is not affected by laterization, but its orthogonal counterpart is.

Leonardo da Vinci, the first to see Camera Obscura as an analogue of the eye, has used it to investigate relationships between the location of models in space and their projections (Fig. 3.10). In his figure projections of the two lateral models are wider than that of the central model. Da Vinci clearly was aware of the effects of lateral displacement. A direct, portable, descendant of the Camera Obscura, the Pinhole Camera, was used by M. H. Pirenne[11] in a number of studies of architectural features, such as colonnades and balustrades, as well of specially created arrays, to analyse the distortions just discussed. The artist is free to use or not to use Leonardo's pane, and if he decides to use it, free to choose how to tackle perceptual embarrassment caused by strict adherence to the rules of perspective.

Subsequent developments of the camera obscura were particularly concerned with elimination of inversions of figures which occurs in the basic form of the camera. Leonardo daVinci describes the phenomenon succinctly thus: "... object which moves in the outer air from east to west will seem in its shadow on the wall ...

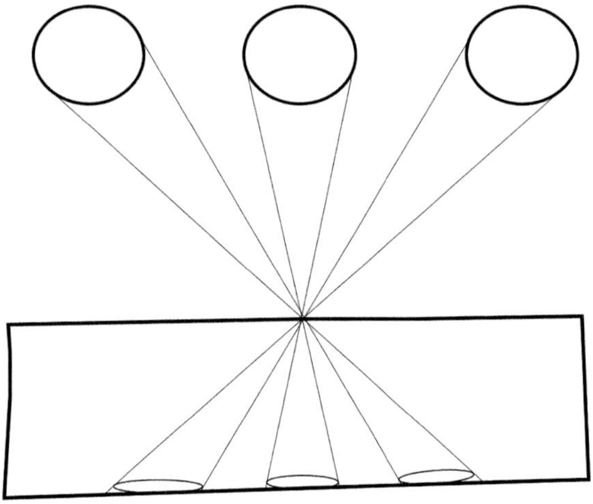

Fig. 3.10 Leonardo da Vinci's camera obscura, the predecessor of Pirenne's pinhole camera (Richter, 1883, facing p. 38). Note that projections of the two lateral rounds are larger than those of the central round. Pirenne has shown that the flatness of the wall at which the images are cast does not question the legitimacy of its comparison to the curved retina of the eye. For development of camera obscura up to its widespread use by painters, see Steadman (2001)

to have opposite motion." (Reich, 1883, p. 76) *Pari passu* an object in the north-west would cast its shadow in south-east. This attribute constituted a major hindrance to the use of camera obscura by artists. Attempts to eliminate it were made by many, as described by Steadman (2001), who researched Vermeer's paintings and found it probable that the great master of perspective resorted to a camera obscura.

Since in the eyes of some use of such optical aids would replace the artist's genius with an unfeeling apparatus, thus reducing his role to that of a skilled copyist rather than a creator, a vigorous debate resulted.

Steadman's conclusion is supported by a Bartelian restitution of the diagonals of the first complete row of floor tiles in Vermeer's *The Lady and the Gentleman at the Virginals.* It was carried out using the reproduction of the picture (Goldscheider, 1958, Plate 38). It reveals no systematic deviations from their true attributes, no steady increase of length with distance from the observer reported by Bartel as present in pictures whose perspectives seem, *prima facie*, correct, but in fact are not correct.

There is, however, the niggling voice of those who "find such paintings (creations of camera obscura) strange and weird because in them form has become content, because their beauty is only skin deep." It may be that the very low esteem of Leonardo's perspective in Bartel's experiment (only four out of 81 students preferred it) accords with this view. It may be, however, that the observed score is less likely to involve aesthetic judgement than perceptual assessment.

Notes

1. See Literary Works by Leonardo Da Vinci compiled and edited from the original manuscripts by Richter (1883): (a) On perspective: Item 83, p. 33, and item 108, p. 64. (b) On camera obscura, item 66, p. 41.
2. Oxford English Dictionary (1998).

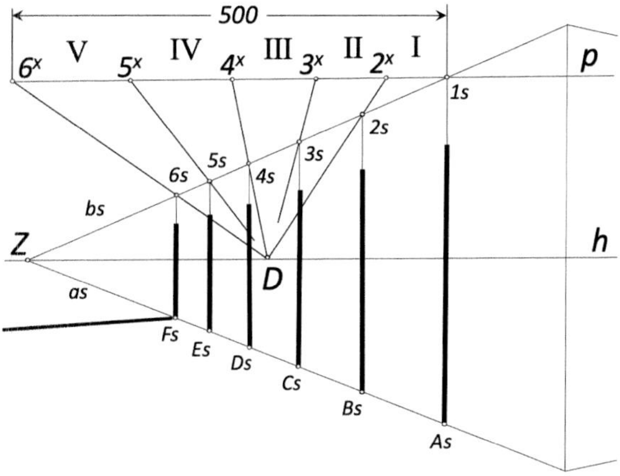

Fig. 3.NI Bartel's restitution of drawn spans (Bartel 1958, Fig. 186)

3. For the relationship between two kinds of cues, see Chap. 2, also Polanyi (1962), Bartel (1958).
4. Kazimierz Bartel (1882–1941) was a university professor in Lwów, Poland. When, in the course of WWII, the German army occupied Lwów, Bartel and 36 other academics were arrested, and 35 of them, including Bartel were murdered (see Deręgowski, 1994; Bartel's life is thoroughly described by a historian Kalbarczyk (2015). Bartel's work on pictures was saved from oblivion by Franciszek Otto, his erstwhile assistant and was published: vol. 1 in 1955, vol. 2 in 1958. The second volume, after considering the principles of human vision, addresses various kinds of perspective and reports investigations referred to here. Of the final chapters, four contain systematic examinations of use of perspective, in antiquity and the Middle Ages, in Italy in the fifteenth and sixteenth centuries, in northern Europe, and in the nineteenth century, and one chapter is concerned with works on perspective published in the sixteenth and seventeenth centuries. Bartel's method of restitution is presented in the sixth chapter. It is briefly described below. The present chapter is largely and shamelessly based on Bartel's far too little-known work. Like de la Gournerie (10) a century earlier, Bartel cut through the cackle by resorting to the laboratory.
5. Bartelian Restitution: Object - to Determine *Relative* Magnitudes of Distances in Real Space from Distances on the picture's Surface. Let the picture show six vertical staves hanging against a wall, as was the case in Bartel's first experiment (Fig. 3.N1). The object is to determine *relative* sizes of the five spans presented in the picture. To achieve this, it is necessary to find in the picture the line of the horizon (h) and locate on this line the vanishing point (Z). This done, proceed as follows. Draw through Z line b intersecting the staves (or their extensions), thus creating six points of intersection 1 s, 2 s, …6 s, of which 1 s is that of the stave nearest to the eye. Through point 1 s draw a line (p) parallel to the horizon (h) and on it mark off a segment 1 s-$6^{\times}$ of an arbitrary, convenient length, say, 500 units, since there are 5 spans. Draw a line 6^{X} - 6 s intersecting the line of horizon at D – the division node. Draw from D lines through 2 s, 3 s, 4 s, … to intersect with line p at 2^{X}, 3^{X}, 4^{X}, … The resulting segments: $1_{s.}2^{X}$, $2^{\times}3^{X}$, $3^{\times}4^{X}$, $4^{\times}5^{X}$, and $5^{\times}6^{X}$ are of restituted relative lengths. Their sum is 500 arbitrary units. (In order to facilitate reference to the original text, the original notation is used.). Bartel also shows that the technique can be employed to determine the proper ratios between depicted spans, given their true ratios, and analyses perspective usage in many schools of art. The grasp of the nature of the construction is facilitated by recalling that: (i) The horizon is infinitely distant, (ii) All parallel lines receding from the eye converge to a point on the horizon, and are therefore depicted by convergent lines. From which it follows that in Fig. 3.NI lines a and b converging to Z define a vertical plane, but lines D-6, D-5 are depictions of parallel lines in the horizontal plane. The two orthogonal planes intersect in the line 6 s, 5 s, …

6. Guczalski (2012) examines an objection to perspective by one of its most vociferous critics (Goodman, 1976) and points out that it is based on lack of understanding of projective geometry, and the fact that parallel lines in a plane parallel to the picture plane project as parallel lines whatever their orientation in the plane, a fact denied by Goodman.
7. See: Deręgowski et al. (1997). They replicated Bartel's investigation using a different procedure. In their experiment the staves were not hung on the wall but mounted on small, electrically driven, trolleys remotely controlled by subjects. The potential effect of the texture gradient of the wall on which the staves were hung in Bartel's experiment was thus eliminated, as was the socially restrictive effect of instructing the laboratory assistant. The incremental trend reported by Bartel from Lwów was also found in Aberdeen, thus the Lwów findings were extended by use of differing samples. Samples used in Aberdeen differed from those used in Lwów, both in gender composition and age. The Aberdeen sample revealed: (i) Significant difference between the sexes, men consistently showing greater distance constancy, i.e. they were less affected by the distances of the judged spans from the observer. An effect possibly related to Witkin's (1961) notion of field dependence; men's spacing was closer to uniform spacing. (ii) Samples of students of architecture at different levels of training did not show that training had any effect on performance of the tasks.
8. See Kubovy (1986) for an erudite presentation of this view. Gombrich (1982) remarks that books on perspective "rarely dwell on the very narrow limits of foveal vision because their topic is geometry, not psychology." See also Deręgowski et al. (1994).
9. Richter (1883). the Literary Works of Leonardo Da Vinci, Vol. I, Item 108, P. 64.
10. See de la Gournerie (1859). The book is available on the web in the Gutenberg collection. His work is notable for the fact that he was probably the first to examine experimentally the perceptual effects of change of a single element in a picture, thus acknowledging the superiority of experimental findings over mere verbiage.
11. See Pirenne (1970) for Extensive Findings Using Pinhole Camera, and scholarly Comment upon them.

References

Bartel K (1958) Perspektywa Malarska. Państwowe Wydawnictwo Naukowe, Warszawa

Deregowski JB (1994) Kazimierz Bartel (1882–1941). Psychologie un Geschichte 5:246–260

Deręgowski JB, Shepherd J, Slaven GA (1997) Sex differences on Bartel's task. British Journal of Psychology 88:637–651

Deręgowski JB, Parker DM, Massironi (1994) Perception of spatial structure with oblique viewing: an explanation of Byzantine perspective. Perception 23: 5.13

Gombrich EH (1982) The image and the eye: further studies in psychology of pictorial representation. Phaidon, Oxford

Guczalski M (2012) Perspektywa. Forma symboliczna czy naturalna? Homini, Kraków

Kalbarczyk, S. (2015) Kazimierz Bartel (1882–19140 Uczony w świecie polityki. Instytut Pamięci Narodowej, Warszawa

Kubovy M (1986) The psychology of perspective and Renaissance art. Cambridge university press, Cambridge

Pirenne NH (1970) Optics, painting and photography. Cambridge University Press, Cambridge

Richter, J.P. (1883), see Leonardo da Vinci

Chapter 4
On Prof. Farish's Immediately Comprehensible and Cezanne's Immediately Puzzling Pictures

Professor William Farish, professor of chemistry concerned with engineering at the University of Cambridge, had a problem. When lecturing he used working models. The models incorporated "loose brass wheels, the teeth of which all fit into one another; axes of various lengths, on any part of which the wheel required may be fixed; bars, clamps and frames; and whatever might be necessary to build up the particular machines which are wanted for a lecture."[1] The models were large, too large to be stored intact. They were therefore dismembered after each lecture. Their component parts were stored and used, as the occasion demanded, either to make different models or to rebuild those dismembered. Since these were working models mimicking the manner of operation of various mechanical systems they had to be assembled carefully with various parts in proper alignment.

Assembly of these devices before each lecture was the responsibility of laboratory assistants. They had to be given precise instructions as to what kind of model was needed. Such instructions could not be verbal (the reader may wish to test his verbal skills by describing the model shown in Fig. 4.1); they could only be visual—a miniature model or an unambiguous drawing showing precisely how and where, in space, each element of the assembly is to be placed. Of the two devices drawing is clearly much more convenient, but of the two well established methods of drawing: the engineering drawing and perspectival drawing, neither proved entirely satisfactory.

An engineering drawing consists of a set of orthogonal projections of a model, such that projections present mutually compatible views and are arranged in a conventional manner which enables the draughtsman to mentally reassemble them and recreate the depicted object, whose solidity is conveyed indirectly.[2] Such drawings are often difficult to read (significantly, the term *read*, not *see*, is used in drawing offices), even for an experienced engineering draughtsman.

Perspective drawings evoke percepts of three-dimensional objects readily, but do so at the cost of direct comparability of dimensions; e.g. decreasing depicted spans between columns or trees, of a line of columns or trees, may show that it extends into the distance, but the depictions of spans cannot be compared directly by

J. B. Deręgowski, *On Picture Making and Picture Seeing*,
Vision, Illusion and Perception 4, https://doi.org/10.1007/978-3-031-23348-7_4

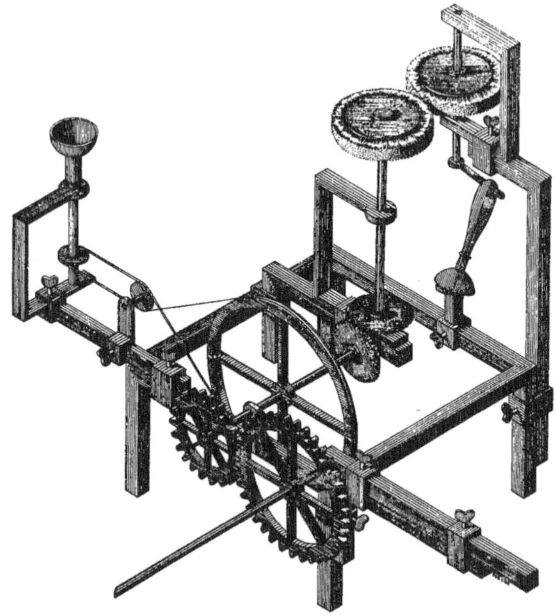

Fig. 4.1 An isometric drawing of the large model used by Prof. Farish to instruct his laboratory assistants. The model illustrates principles of power transmission (From Farish 1822)

measurement. This may not concern a gallery visitor but would concern a laboratory assistant who is intent on precise spacing.

To help his assistants Farish developed a new system of depiction, such that pictures drawn in accordance with its rules look three-dimensional and are also scalable. He called his system isometric perspective (it is now generally referred to as isometric drawing). The kind of problem the laboratory assistants had to tackle is shown in Fig. 4.1. To the laboratory assistants' great relief, the drawing shows unmistakably a three-dimensional machine. To their even greater relief all the distances parallel to the principal axes (the vertical and the other two inclined at 60° to the vertical) are commensurable. Therefore, the position of the inner bracket supporting the vertical axle can be read from the drawing. They also noticed that the angles at which axes and lines parallel to the axes intersect *look* like right angles; as does, indeed, the angle between the axes of the arms of the cowgirl (Fig. 2.4b, c) who instead of attending to her job plays with gossamer (True angle 130°). It appears, therefore, that not only do oblique lines evoke the perception of depth, as shown earlier, but also that such lines when intersecting affect perceived space in a very distinct manner.

Isometric drawings which present so meagre depth cues and fall so short of Leonardo's notion of perspective, but which Farish's laboratory assistants found more helpful than perspective drawings, question the cliché: perspective yields the most realistic images.

Observations of people from a sub-culture in which pictures play but a minor part are in agreement with these findings, and support the notion that distinction should

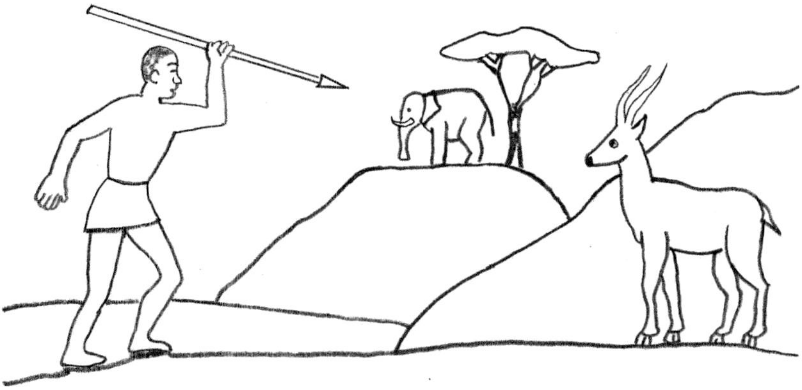

Fig. 4.2 A test picture from Hudson's Test, incorporating the cues of familiar size (elephants are larger than antelopes and men) and pictorial overlap (the antelope overlaps a hill which overlaps the hill on which the elephant stands; the man overlaps a field which overlaps the hill on which the elephant stands). The questions which the subject had to answer concerned distances between the hunter and each of the two animals, such as the question: "What is closer to the man, the elephant or the duiker?" The response "elephant" would show absence of pictorial depth. Hudson's test booklet consists of seven figures, exploring various pictorial depth cues (Reproduced with W. Hudson's kind permission)

be made between perception of pictorial cues associated with daily experience such as relative size, elevation, clarity of view and cues responsible for perception of depth in isometric drawings. Hudson's Pictorial Perception Test and the Construction Test explore this issue. They differ radically in the cues which they embody. The Pictorial Perception Test consists of a series of pictures (Fig. 4.2) presenting various combinations of traditional depth cues and, in their context, figures of a hunter, an elephant, and a duiker. The test's score is determined by recognition of depicted spatial relationships among these three key figures. The Construction Test requires that models shown in simple geometric drawings (Fig. 4.3) be built; built models are judged as either essentially two- or three-dimensional. Many of the subjects from a non-pictorial culture who failed to see depth in Hudson's pictures nevertheless saw it in the geometrical drawings of the Construction Test, in spite of the fact that the drawings did not represent anything familiar. Such ease of perception of depth with purely geometric drawings is also found when people are asked to set a pair of wooden calipers at an angle matching that shown in a picture. This they can do either by attending to the angle measurable in the plane of the paper (thus not responding to pictorial depth) or responding to the depth cues inherent in the picture of that angle. There is a general tendency to attend to the depth cues.

Pictorial depth, it appears, is perceived in linear, non-perspectival pictures by populations with very meagre pictorial exposure; they respond to pictures of objects

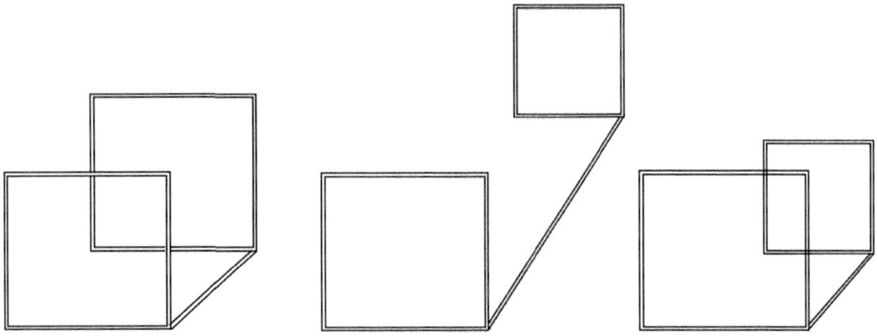

Fig. 4.3 Three of the drawings used in the Construction Test. Subjects built models using bamboo splints and plasticine. The models built were judged to be planar or spatial

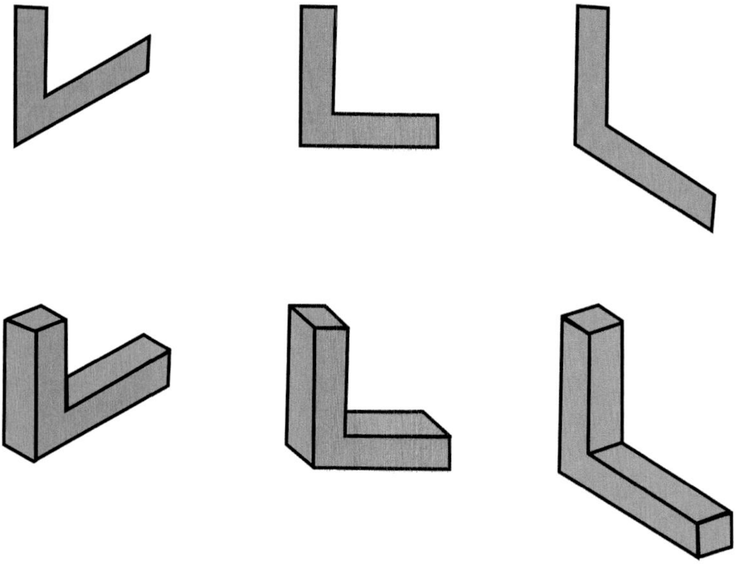

Fig. 4.4 Kwengo Calipers Test. Subjects set wooden calipers at the angle they saw in the picture. The pictures were either mere outlines (top row) or presented depth cues (bottom row). Comparisons of responses to the corresponding figures in the two rows would in the case of perceivers of pictorial depth, be: first column, $b > a$; second column, $b = a$; third column, $b < a$

with which they are not familiar, by means of a novel device (the wooden calipers) in circumstances which they had never experienced before (Fig. 4.4).[3]

It was shown in Chap. 2 that an isolated sloping line evokes perception of depth and that a compact of lines, whether derived from a meaningful entity (such as a painting by Vermeer or Chełmoński) or of abstract origin (as in the case of the illusion figures) creates a perceptually coherent, spatial, three-dimensional percept;

this perceptual effect may be, it appears, influenced by environmental exposure but seems in its essence, universal.

There is a genre of visual stimuli beloved by students of perception, involving these phenomena. They are known sometimes as "impossible objects" and sometimes as "impossible pictures". These are clear linear drawings which are such that the perceptual mechanism, obedient to its inherent purpose of making sense of the environment, in spite of all its efforts cannot make sense of them. These drawings do not present overly strong traditional depth cues, yet their percepts are pronouncedly spatial and ambiguous, and their ambiguity is spatial, so that now one, now another spatial version prevails. An example of such a drawing is shown in Fig. 1.10 which shows a two-pronged trident. The ambiguity it presents lies in the number of tines—are there two square or three round prongs? As noted in Chap. 1 the lines connecting the two ends of the figure change their perceptual significance in the course of their travel from the two-pronged to the three-pronged end. One would expect the desire to remember such an unstable figure to result in suppression of the confusing spatial effect, and seeing the figure as flat. This does not generally happen, and for this reason subjects cannot recall such figures easily and find reproduction of impossible figures from memory difficult.

When a drawing is attempted of an impossible figure continuously on display, the figure can be drawn reasonably easily, copying it line by line, but when the figure is shown only for a short time and an interval of time is allowed to lapse before drawing begins the difficulties arise and grow rapidly with the duration of the interval. An interval of ten seconds instils a considerable obstacle. The reader can experience these difficulties directly. Get a pencil and a piece of paper, cover Fig. 1.10b, c, and look at the three-pronged trident (Fig. 1.10a) and very slowly count to ten. Then, close the book, wait for ten seconds, and attempt to draw a facsimile of the picture that you have seen. If the drawing made matches the one in the book you have a rare ability and are a better man than the author. If it does not, a much more likely result, repeat the procedure. Repeat it as often as you will.

The tenacity of the incongruent three-dimensional percept is the evidence of the eagerness of the perceptual system to see depth. It perceives depth not *pictorial* depth. The adjective pictorial is a classification term used by researchers. It perceives depth because, just as in the case of perceiving objects, it is responding to cues present in the natural environment.

The subtlety that is at the root of isometric drawings, is exploited by some artists to create compellingly convincing depictions of space. Such is the architectural sketch in Fig. 4.5a.[4] Bartelian restitution of the left wall of this sketch (Fig. 4.5b) shows that the chamber constructed on assumption of its perspectival correctness would differ greatly from the perceived chamber. These pictures do not merely deviate from the rules of perspective whenever strict adherence to them would prove perceptually irksome, they ignore these rules entirely. Such deviations from the rules of perspective acknowledge both the usefulness of Leonardo da Vinci's perspective and its limitations. Professor Farish's isometric perspective shows that another, and in the view of his laboratory assistants' superior, way of depicting objects exists. Impossible figures are seen as impossible because they are seen as having three

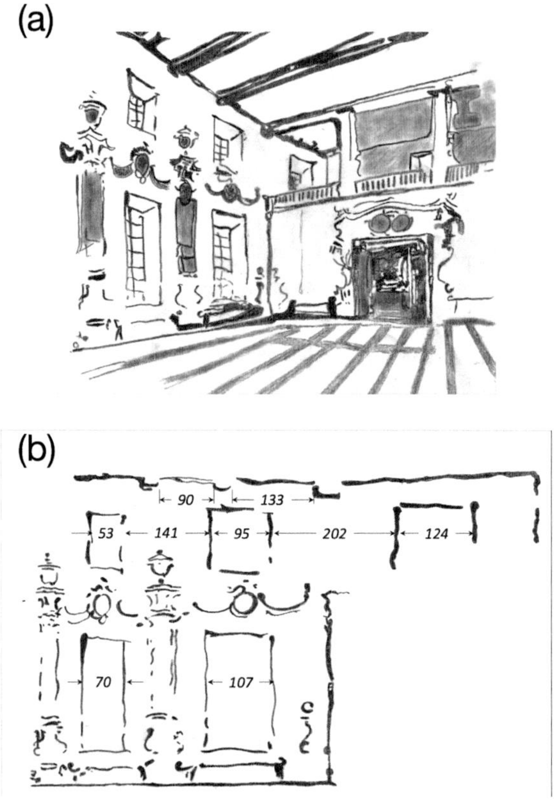

Fig. 4.5 S. Noakowski's Sketch of a room. **a** Original sketch (Derived from Bartel, 1958, Fig. 183). **b** Restituted dimensions of the left side wall (Derived from Bartel, 1958, Fig. 184) Of the three small windows at the top the nearest to the viewer, appears somewhat larger than the furthest, but on restitution it is shown to be less than half its size. A portfolio of sketches by Noakowski (1953) presents 20 examples of his style

dimensions; a picture *seen* as an *arrangement of lines on a flat surface* cannot appear perceptually impossible. Pictures composed of recognizable, but cognitively incongruent elements, such as birds' wings and the body of a horse, such as Pegasus, do not cause perceptual puzzlement in the way of impossible figures. The monopoly of classical perspective is thus made uncertain. Another assault on this monopoly derives from by Byzantine Perspective.

Inverse (also called Byzantine) perspective is routine in the art of many cultures. In 'Byzantine' depictions the lines representing parallel contours extending away from the viewer do not converge.

It is particularly noticeable when features known to be rectangular are represented by trapezia whose parallel edges are intended to be seen as at different distances from the viewer and therefore, in accord with the accepted pictorial depth cues, edges further away are drawn higher up in the picture's plane, *but contrary* to such cues they are also drawn longer. Pieces of furniture, oriental garden pools and rugs are thus portrayed. The table in a Russian icon is drawn in Byzantine perspective (Fig. 4.6) as is the royal bed in a West European medieval miniature (Fig. 2.N3) and the seat in the Chinese courtly garden portrayed by Gu Kai Zhi (344–406) (Fig. 4.7).

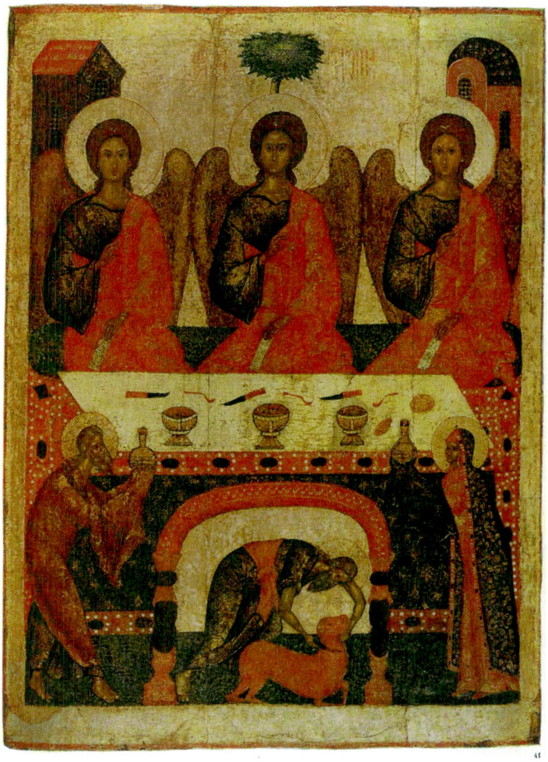

Fig. 4.6 Pskov icon (early 16th cent.). The table is drawn in inverse perspective and the three chalices, which are presumably solids of revolution, show their uppermost edges approximately elliptical and their bases as rectilinear. This is an extreme case of the edge of the base of a chalice being shown as forming a slimmer ellipse than the rim of the cup. The coffee pot on Cezanne's Kitchen Table (Figs. 4.15 and 4.N1) shows the same disparity but in a less drastic form

This painting is of particular interest, because Chinese artists of that time painted in a manner entirely different from that of their Western counterparts. "Painting was generally executed with a speed inconceivable in the West—sometimes in a few seconds—but much thought preceded it. The conception was completely present in the mind before the artist began work and execution was automatic and spontaneous. No corrections were possible" (Oxford Companion to Art 1970, p. 230). The notion of inverse perspective in the painting was therefore of the mind, not of immediate visual experience; it was certainly not thought of as a stigma of incompetence.

Sometimes oriental pictures, notably Persian miniatures, show the compromise between convergent and divergent perspective (or, perhaps, abandonment of any notion of perspective); square objects are represented by squares. This is so in the case of the rug on which the sage physician and the ailing maid are seated in Fig. 1.15.

Inverse perspective is sometimes dismissed as a result of a lapse on the artist's part of no relevance to the *modus operandi* of the perceptual system, or as a deliberate distortion intended to convey information beyond immediate apprehension of the perceptual system. Such claims may on some rare occasions be justified, but each of them would require supporting evidence because empirical evidence that inverse perspective is a naturally occurring phenomenon is abundant.

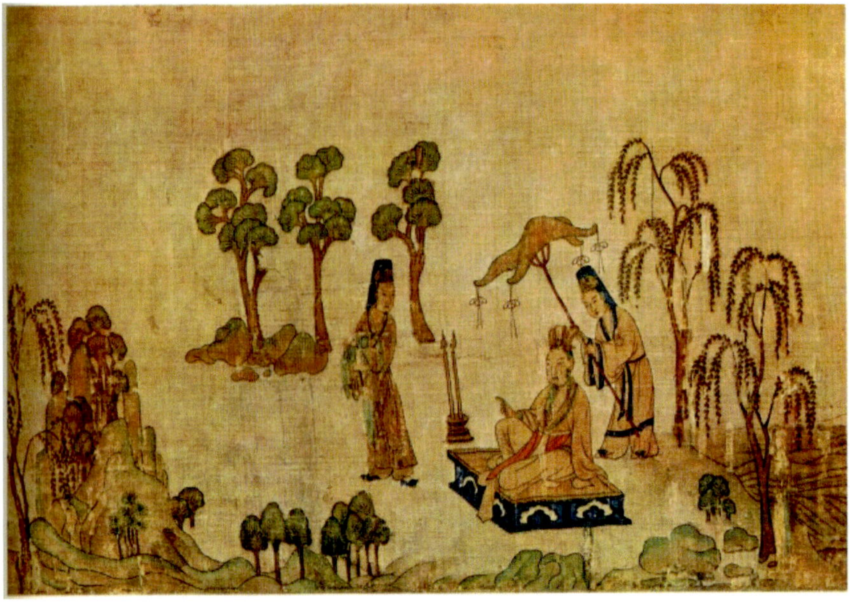

Fig. 4.7 "A Nymph of the Luo River" by Gu Kai Zhi (China, 344–406). Note the shape of the seat, which is presumably rectangular and drawn in inverse perspective

Inverse perspective was observed in an experiment in which observers seated in a darkened room faced two parallel luminous tubes extending away from them, one slightly to the left and the other slightly to the right (see Fig. 4.8, tube A–B and tube C–D). They saw imaginary extensions of these two parallel tubes as having a distant point of convergence, just as Leonardo da Vinci's perspective dictates. However, when a third parallel, but laterally displaced, tube was added (tube E–F in the figure) it was seen as parallel to the original pair, but not as pointing at their point of convergence. When tilted so as to share the point of convergence (position E'–F in the figure) it was no longer seen as parallel to the original pair.[5] The laterally displaced tube's extension appears to intersect the observer's median plane at a point more distant from the observer than the imaginary intersection of the original pair. Therefore percepts of the pair CD:EF are divergent. Quadrilateral ABDC, as seen, complies with the rules of convergent perspective; quadrilateral CDFE does not, it is a trapezium.

A later experiment using luminous points, described in the Chap. 2, confirmed this finding. Both these experiments required subjects to respond to stimuli in the horizontal plane. Further evidence that inverse perspective is seen when three-dimensional objects are viewed is provided by yet another experiment using luminous stimuli in a darkened laboratory.[6] Luminous rods were so arranged that they appeared to form edges of a box. Remote controls made it possible to vary angles among the rods and therefore to vary the perceived shape. Subjects used joysticks to adjust spatial orientation of the rods so as to create 'a proper rectangular box'.

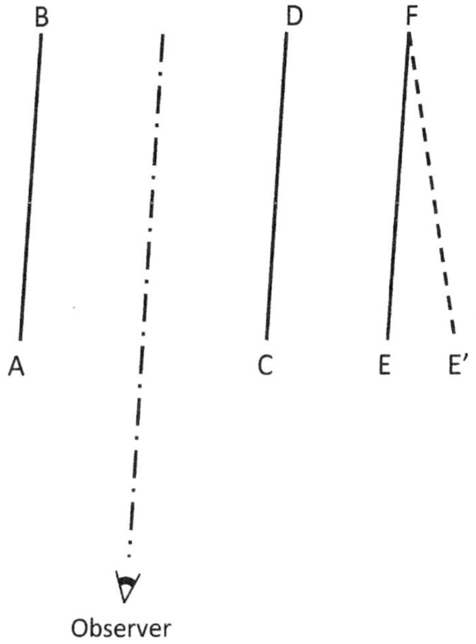

Fig. 4.8 Arrangement of luminous tubes in ten Doeschatte and Kylstra's (1956) experiment. Plan view: A–B, C–D and E–F are luminous tubes. Tube E–F had to be set at E'–F to appear to converge the same point as the two other tubes. Tubes seen as parallel were not seen as convergent but as divergent. This questions the psychological validity of pictures painted on Leonardo's pane, using geometrical perspective. It is noteworthy that Byzantine depiction of further elements as larger than physically equal, nearer elements is analogous to percepts of monocular inversion wherein physically closer elements are seen as further away and larger than elements physically further (Deręgowski 2014)

They did so completely unrestrained in their head movements using binocular vision from several stances, in circumstances, therefore, identical with those experienced daily. When the 'box' was presented in front of the subjects, and would therefore appear convergent with distance, the rods were set as divergent. When the 'box' was moved sideways, the rods were set as convergent. The results, just as the results of the 'planar' investigations, showed that inverse perspective is experienced in laterally displaced vision, that therefore it is neither an aberration nor a figment of the artist's imagination. Artists can choose from the gamut of their visual experiences; the manner in which they choose is, of course, their own. The choice may be met with approval or disparagement by the cultural milieu.[7]

Pictures in the Byzantine perspective differ from those in convergent perspective of Leonardo's window, not only in their gross appearance, but also in subtler qualities of representation. The Ponzo illusion reveals this, generally overlooked, difference.

Presented in its usual manner the Ponzo figure (Fig. 4.9a) shows two oblique lines converging upwards and two parallel and equal short segments (the bars) between these lines. The implicit assumption which the viewer of the figure makes is the assumption generally made when viewing pictures—that the bottom of the figure is perceptually closer to the viewer. The illusion affects the perception of the relative

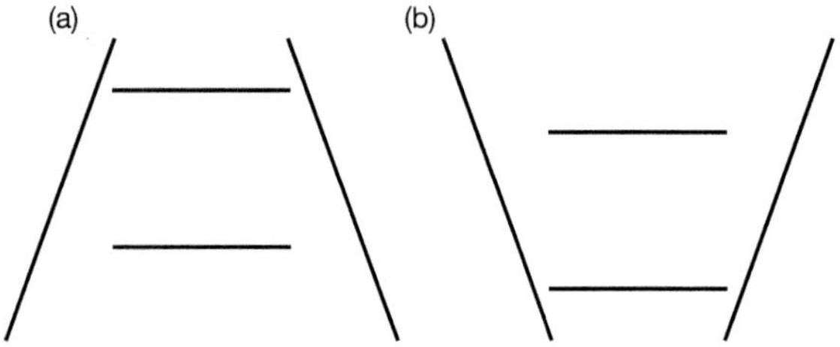

Fig. 4.9 Two orientations of the Ponzo figure: **a** as usually presented; **b** inverted. The illusion, that the bar nearer to the narrow end is larger, is greater in **a** than in **b**

sizes of the two bars. Two vectors are responsible for the perceived sizes: (i) perspectival convergence, (ii) perceived closeness to the viewer (see Fig. 2.1). Both vectors enlarge the upper and perceptually more distant bar. Compounding of these vectors determines the percept—the upper bar is seen as larger than the lower bar. When the figure is turned upside down (rotated by 180°) (Fig. 4.9b), the two vectors are no longer consonant; the convergence vector affects the lower bar, but the perceptual depth vector affects the upper bar. The illusion is therefore weakened, and the two bars appear to be more similar than they do in Fig. 4.9a. The Byzantine perspective, it appears, is a worthy competitor with the perspective of Leonardo's window.

The Byzantine perspective has, it seems, a respectable home in lateral vision, whence it can be brought out as a perceptual compromise, a device much affected by the artist's experience. Primo Zeuxis, intent on presenting bisons *as they are*, by their typical contours is not prone to such experiments. There is, however, a family of models which has provoked numerous artists, in pursuit of pictorial space worldwide to distort typical contours. Depictions, throughout centuries of art, in diverse cultures, of bowls, chalices, candlesticks, wine glasses, and other items of the geometrical family of solids of revolution illustrate this admirably. These depictions reveal a prevailing tendency. The ellipses depicting fully exposed tops of the models are more rotund than are the half-hidden ellipses of their bases. The latter are even occasionally found to be but straight lines (as shown in Fig. 4.6).

Since the described tendency appears to be independent of cultural milieu it probably originates from the manner in which the perceptual mechanism operates. This hypothesis is confirmed by freehand drawings made by Scottish students of engineering. They drew a small cylinder standing on the table. They drew, as did many artists of many cultures before, a cylinder whose upper ellipse approximated to the circle more closely than the ellipse of its base. When the cylinder was replaced by a model consisting of two parallel circular lamellae connected by thin wire, thus mimicking top and bottom of the cylinder, the drawings they made showed no such trend.[8]

Drawings of the cylinder show that the conceptual planes of its top and bottom surfaces are not parallel (as they physically are), for parallels surfaces would be represented by similar ellipses. As a rule, however, the lower face of the cylinder is drawn as a narrower ellipse. This is the effect which would prevail if horizontal discs at different distances from the observer were drawn; the more distant disc would be portrayed by a narrower ellipse. The hypothesis is that the depictions of the cylinder obtained results from treatment of the vertical distance as horizontal (a vector contributing to the horizontal-vertical illusion, and the Ponzo illusions, *q.v.*) and therefore the phenomenon is allied to that of Byzantine perspective.

Consider the Pskov ikon (Fig. 4.6). The bases of the chalices are shown by straight lines and, by implication, lie in the horizontal plane, but the rims of the cups are shown with their more distant halves above the less distant, which implies that they lie in an inclined plane which diverges from the horizontal plane with increase of distance from the viewer.

If such portrayal of goblets, chalices, cups, and other utensils, so common in art, is indeed a muted demonstration of inverse perspective, then treatment of inverted perspective as a cultural aberration not worthy of the attention of students of vision, seems mistaken.

Responding to both models the students drew what they saw and they saw the two edges (top and bottom) of the cylinder as of different shape but the edges of the lamellar discs as of the same. These differences of percepts came about because physically identical curves of the top and bottom surfaces of the two figures were seen in different contexts. Different visual cues were involved in generation of the percepts and so the percepts were different.

Such perceptual spatial effects, it appears from their meticulous execution, were deliberately used by the Bushman artist, living in a highly sophisticated pictorial culture, in petroglyphs of bovine heads. Some of these petroglyphs show a head with one of the horns split transversely (Fig. 4.10a). This split cannot be reasonably attributed, either to wear and tear of time (the petroglyphs are well preserved) or to the artist's carelessness (the cuts are very neat). The artist, one has to conclude, has deliberately created the gap. Its purpose is exposed by comparing perceived headings of depictions showing such a gap with their modified versions in which the gap is obliterated (Fig. 4.10b). Such comparison discloses that the gap affects the perceived headings of the depicted objects and suggests that it might have been introduced for this very purpose.[9] Used to modify pictorial depth, it serves as a little-known partner of symmetry.

Symmetry affects pictorial depth in a very complex manner because the same objects may be seen as symmetrical or asymmetrical depending on the point of view.

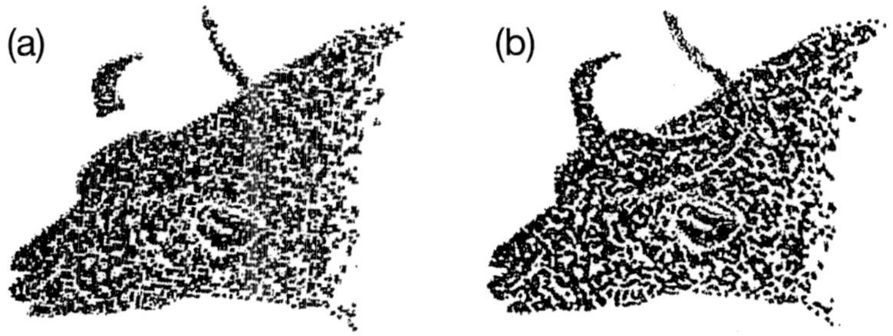

Fig. 4.10 Pictures of Bushman petroglyphs. **a** Petroglyph, as found. **b** The same picture with the gap filled. Filling the gap affects perceived direction in which the snout is pointing

The bison (of Chap. 1) whose picture Zeuxis Primus intends to make can project a symmetrical pattern, as can most animals. Its frontal view and its rear-view project symmetrical images, but both these images have the peripheral contour which is far less salient than the asymmetrical peripheral contour of its side view. Zeuxis wants to create a recognizable picture, not an optical puzzle. He chooses therefore as the core of his picture the most salient contour of all, the typical contour, the asymmetrical side view. There are, admittedly, parts of the animal, such as its horns, whose typical contours may lie in different planes than that of the typical contour of the body, but these are but adjuncts and can be treated accordingly. In the cave of Altamira, horns, as of lesser importance, are sometimes represented by single lines (Fig. 1.5a). In Lascaux[10] the largest group of bovids have horns which form pairs of about identical and identically disposed elements, such as presented by two bananas laid side by side. None of the Lascaux horns approach the splendor of the symmetrical crown of horns of the Heavenly Cow of Egypt (Fig. 1.18).

So dismissive a treatment of symmetry does not imply that the Paleolithic artist was unaware of the symmetry of objects or that he did not perceive it in a way his modern counterpart does. It only means that symmetry was not fit for the purpose when making representational pictures. The purpose for which it was fit was that of satisfying aesthetic yearnings by means of abstract designs such as band patterns. Such artistic creations are free of any obligation to attend to the third dimension. Symmetry is ubiquitous. An extensive cross-cultural study[11] supports this view. It shows that symmetrical elements are greatly favoured in most of the cultures, and such elements are overwhelmingly symmetrical about the vertical axis.[12] Thus the concept of symmetry 'the great organising concept'[13] of sciences and mathematics served, for a time, before assuming such an exalted role, as one of the central aesthetic concepts. The perceptual mechanism always strives to make sense of the entire visual field and in so doing it makes use of regularities which the vista displays. Symmetry is such a perceptual regularity. Early encounters with literacy are also encounters with symmetry. A primary school pupil asked to copy 'cab' may respond with cad, sometimes with cap and sometimes even with caq. This bane of early schooldays is shown in Fig. 4.11 which defines the network of relationships among the four elements b, d, p, and q.

The travails experienced by young writers imply that 'b' is not perceptually encoded only as 'b', but also as 'd', 'p', and 'q', and that the strength of those derivative encodements declines in the order stated. The most common and enduring affinity is that between b and d. It is expressed not only by mistaking one for the other, but also by preference to set bilaterally symmetrical patterns with their axes of symmetry in the vertical plane. The effect, it has been postulated,[14] serves to encode the spatial orientation of stimuli.

A picture in Emperor Henry II's Book of Pericopes (Fig. 4.12) shows men rising from their graves summoned by angles' horns. Each grave is depicted by the trapezium of its top, the trapezium of one of its sides, and the rectangle of its front face. Both trapezia have their longer parallel sides uppermost in the picture plane, thus creating a particularly telling display of use of inverse perspective to depict orthogonal surfaces. The graves are drawn as the artist remembered them, or perhaps

b d

p q

Fig. 4.11 Arrangement of four letters showing their symmetrical relationships, with respect to: (i) vertical axes of symmetry: b–d, p–q; (ii) horizontal axes of symmetry: b–p, d–q; (iiic) polar axes of symmetry: b–q, d–p

as he remembered other paintings of graves. The graves are placed in no particular order in the quadrilateral graveyard, at the corners of which angels stand. Because there are no pronounced texture gradients such as are seen by Farmer McDonald and Doddie Munro (of Chap. 2) gazing at a field of cabbages, or looking down a street, the graves can be seen as either on a level burial ground or as floating in the air stacked in two rather untidy columns. The pictorial depth cue is that of elevation—the further away an element lies on the horizontal plane below the viewer's eyes, the higher it is shown in the picture. This does not abolish the possibility that the elements shown higher in the picture may have been so depicted because the artist wished them to be seen as higher up rather than further away. Whether the artist used this cue of elevation is a matter of speculation. The picture presents no supporting evidence, elevation is the only potential depth cue used and other potential depth cues are blatantly ignored. The two angels who according to the elevation cue are furthest away are of the same size as the two angels who are nearest. The resurrected likewise do not decline in size as they, supposedly, become more distant, which disposes of the argument that the angelic violation of the cue merely shows that the heavenly creatures are not subject to earthly rules of perspective.

The pictorial depth cues already described (Chap. 2) are direct and rely solely on the imitative power of pictures; the greater the optical similarity between the stimulation provided by the picture and that provided by the real-world model, the stronger are the cues with which the artist imbued the picture. These imitative cues do not expressly acknowledge involvement of the perceptual system in forming the percepts. The system can be regarded as steadily, impartially performing its function of mirroring. However, when a picture does not match the real world the manner of involvement of the system can be questioned. This is so in the case of El Greco, whose elongated figures caused some of his contemporaries and some of the succeeding generations, to question the health of his eyes, specifically that he suffered from astigmatism, and even his mental wellbeing.[15] If correct, then such

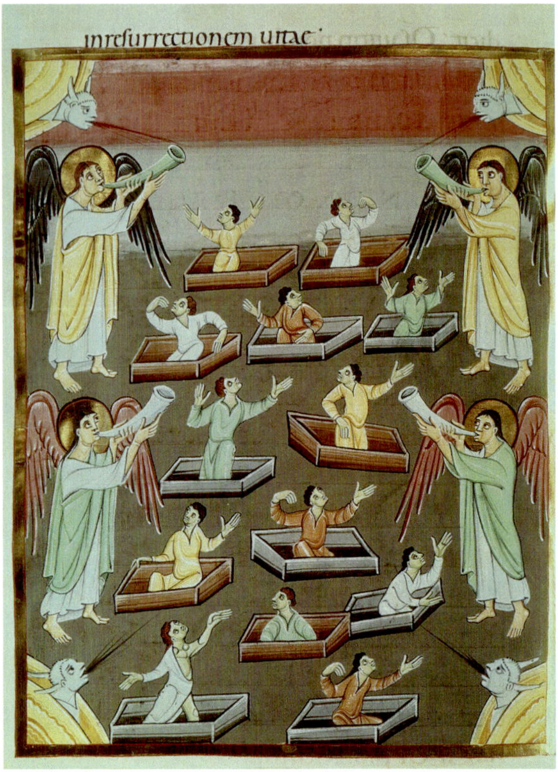

Fig. 4.12 The graves opening on Judgement Day are all shown, by the miniaturist of the Book of Pericopes of Henry II (11th cent.), in Byzantine perspective

afflictions should, *pari passu,* be attributed to other outstanding artists; Theophanes, the Greek (Feofan Grek, c. 1370–c. 1405), the icon painter whose portrait of St Paul is juxtaposed with El Greco's portrait of St Andrew in Fig. 4.13, to Parmigianino (1503–1540), whose works El Greco saw and admired in Rome as well as to numerous other artists: all produced distorted representations. It is, however, unlikely to be correct. El Greco's work itself provides weighty arguments against this notion. There is the weighty ophthalmological evidence succinctly presented by Marmor and Ravin[15, 16] and there is also the simple truism[16] that in order to evoke a percept identical with that evoked by the stimulus one should reproduce the stimulus, not its percept. Elongated figures, figures showing forelengthening, occur, as has been shown on Leonardo da Vinci's screen, they are evoked whenever the artist paints a model, not as it is seen when looking at it directly, but as it would appear, on the screen when presented away from the line of sight. When the displacement is lateral, lateral extensions occur; when it is vertical the extensions are vertical, and so radially in every direction of displacement from the point at which the line of sight strikes Leonardo da Vinci's pane.[17] In consequence, depiction of a model placed away from the line of sight is expanded radially and when mistakenly taken to be a depiction of the model as seen in central vision, is deemed to be distorted. Strzemiński[18] proposed that this effect laid the perceptual foundations of Cezanne's style. In Cezanne's picture of an

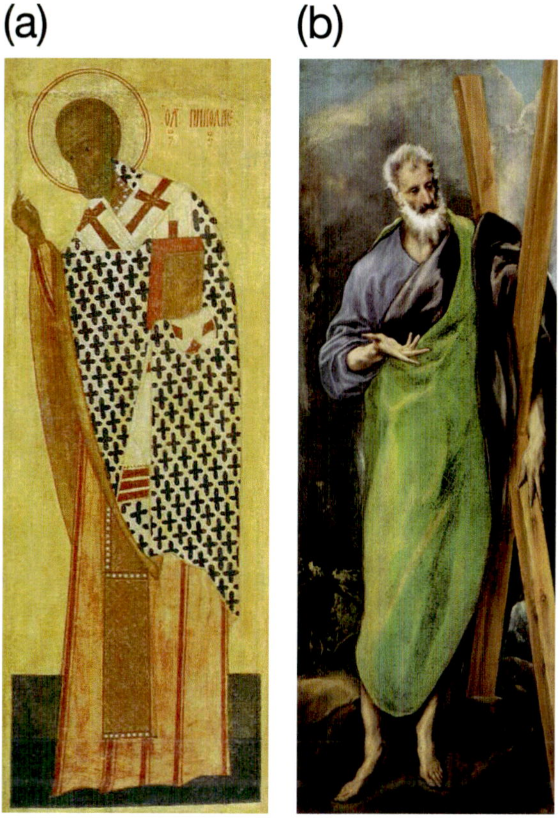

Fig. 4.13 Very tall and slim saints. Saint Paul on an icon by Theophenes the Greek (14th cent.) and St. Andrew by El Greco (16th cent.)

overburdened table (Fig. 4.14) various, we expect, in fact symmetrical utensils seem distorted; the tureen, the coffee jug, and the large pot are asymmetrical. Asymmetry of each of them would be expected if the artist painting each of them fixated, not on its axis of symmetry, but away from it. If he did so, then parts of the peripheral contour further away from his line of sight would be more affected than those near it. Asymmetry of the resulting percept would cause the abolition of the axis of symmetry but may hint at a central line. Such line would generally be sloping as it, very pronouncedly, is in the case of the coffee pot and tureen. Such changes may, this suggests, often be accounted for by distortions of the peripheral contours. In Fig. 4.15 a square balanced on one of its corners (A) demonstrates how such transformations of peripheral contours take place. Corners B and D of the square are equidistant from its vertical axis of symmetry, passing through corner C (BC = CD), but projection of such a square on Leonardo's window normal to the line of sight is not symmetrical (bc < cd). A projection seen in a side view renders the square as a deltoid (See notes[17] and[18] for further discussion of this issue).

There must, one suspects, be a perceptual reason for placement of the artist's line of sight at a particular spot of the model—an act greatly affecting portrayal.

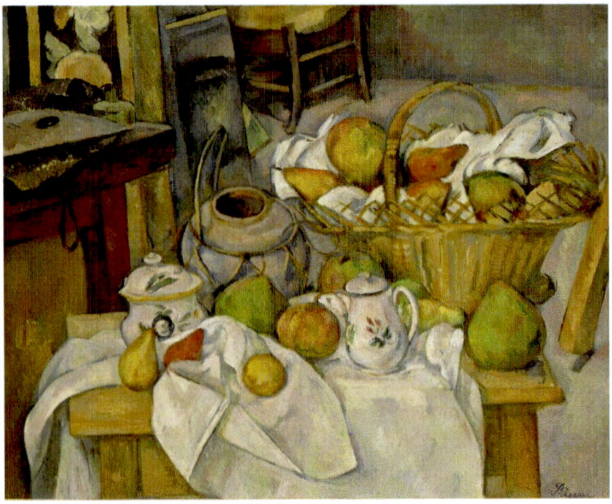

Fig. 4.14 P. Cezanne's *Kitchen Table*. Note the asymmetry of the depictions of utensils which are assumed to be symmetrical

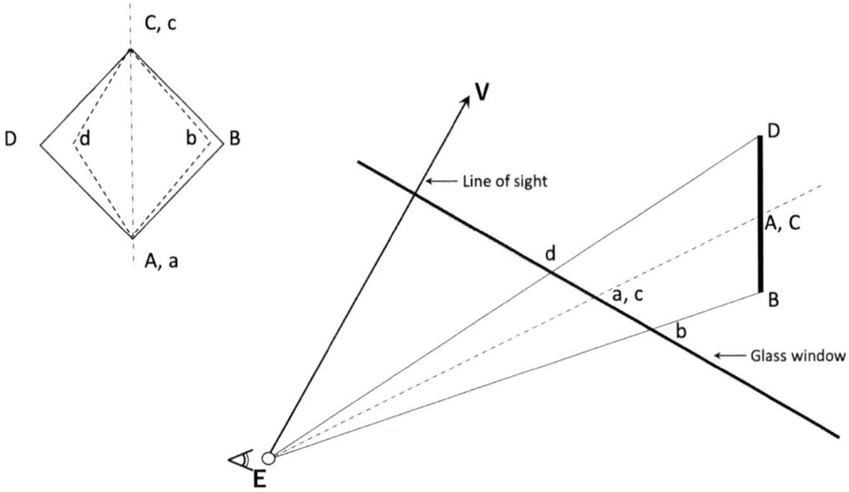

Fig. 4.15 A bird's eye view of the observer's eye (E) looking in the direction V. Its visual plane is perpendicular to its visual axis EV. A vertical square ABCD, resting on corner A, projects on the plane as a quadrilateral adcb

Strzemiński argues that the locations on the model at which the eye rests are the visually salient points. To the left of the tureen there is an area of intense linear and colour contrast; another such focus of contrasts (of colour, of shade and linear) is found near the apple to the left of the coffee jug; yet another similar focus is marked near the orifice of the large pot, denoted by the colour contrast between the orifice and the outer surface, and linear contrast of a straight upright handle and the roundness of the lip. Cezanne's table on which these utensils rest presents yet another visual conundrum—its top is discontinuous; it is split into two distinct planes and therefore implies two distinct orientations of the artist's line of sight. The meeting line of the two discrete parts, which would reveal this embarrassing fact is discreetly hidden by kitchen cloths overhanging most of the edge. The two orientations of the artist's line of gaze were such that, according to Strzemiński, when looking towards the left-hand corner of the table the gaze fell on the table's leg, so that a large expanse of the table top was in the peripheral visual field, but when looking towards the right hand corner the gaze fell on the table's top and an expanse of the table's leg was in the peripheral viewing field. The settings of the lines of sight are such because on the left there is an intense light contrast just below the table top on the table's leg, and another high contrast area among the fruit piled near the edge on the right-hand side.

The perceptual process involved derives from the viewer's natural tendency; 'viewers normally feel that their mind's eye is on the perpendicular to the picture plane'.[19] In Cezanne's case the process was reversed. There was no picture plane in his mind's eye, but there were salient points within his model, and these determined the direction in which his eye looked; this in turn determined the inclination of the projection plane and so produced the pictorial distortions. The act of painting was that of transferring projections on such a notional plane determined by the direction of gaze onto the real plane of the canvas.

There is a similarity between Cezanne's task and that of Zeuxis Primus. Zeuxis Primus takes with him into the darkness of the cave memories of bisons that he has seen with their typical contours in his fronto-parallel plane, the plane normal to his line of sight. His memories are composite—he has seen many bisons in various circumstances. He intends to transfer the bison onto the plane of the rock face. Likewise, Cezanne transfers the contours determined by his line of sight on to the canvas's surface, but when he takes different lines of sight, he incorporates the associated distortions into the same picture. An apparently, but only apparently, similar argument could be sought to explain the manner in which the Kakadu crocodile and others of its kind are drawn. Such an argument would rest on a crude misapprehension of the difference between the two styles. The Australian picture does not attempt to convey the artist's subtle experiences. It is composed of typical contours honestly displayed. The Australian artist does not attempt to cover embarrassing junctions and paints from memory.

The European paints from a carefully arranged model. When painting the table, the eye of the artist rested on the specific focal point and his hand drew what he saw. The depiction of the overburdened table is not that of a table as it was, it is not a map, it is of a table as seen by the artist whose eye was mobile and who compounded different views. Thus the Kitchen Table presents in a single plane and amidst spatially inert depictions of objects lacking clearly defined contours, such as the tablecloth and the fruits, spatially inconsistent views of the three utensils. These utensils appear to be painted from three distinct stances and portrayed as they would appear in planes normal to those differing lines of sight. The direction of each of the three lines of sight is such that a uniform utensil next to it appears malformed. A tacit assumption prevails that the utensils were of regular shape but there is no irrefutable evidence to

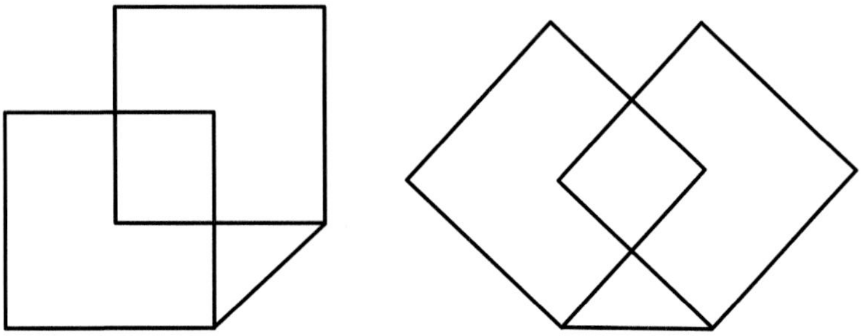

Fig. 4.16 Visual effect of symmetry about the vertical axis demonstrated. The figure on the left is more likely to evoke perception of space

support this reasonable assumption. It is possible to speculate, also without adducing any supporting evidence, that the utensils on the table were in fact malformed and therefore the picture, painted from a single stance shows pieces of crockery, bought by Cezanne because he found their deformities interesting.

Such explanation, however, fits ill with the split table with which Strzemiński's explanation deals with ease. Strzemiński's notion therefore stands.[20]

A visitor to a museum looking at the picture or a person viewing its reproduction in a book sees the utensils as malformed. He sees them thus, not because of extensive experience with regularly formed utensils (although this might influence him), but because the shapes of the utensils approximate to symmetrical shapes and their axes are nearly vertical. When presented with stimuli (say, arrays of small tiles) embodying the characteristics of near symmetry of the pattern or/and of near verticality of the axis and asked to reproduce them, subjects tend to err towards symmetry and towards verticality of the axes[12]. A more immediate, but convincing, demonstration of the effect of symmetry can be had by looking at Fig. 4.16 with the book held in front, then turning the book through 45° clockwise. This will change the figure's appearance from that of two squares separate in space to that of a symmetrical, planar arrangement of the two squares, connected by a horizontal line.

The eye can not only point in a particular direction, it can also bestow its attention on objects at a particular distance—such objects are seen more distinctly than others. The line of sight having both direction and magnitude is therefore a vector, and the artist is free to record changes of its two qualities, and thus to record wanderings of the eye.

Two factors determine the lucidity with which any element of a scene is seen: the direction of the line of sight and the accommodation of the eye to distance. The elements which are in the central visual field and at the distance accommodated to are seen most clearly. Elements which are in the peripheral visual field and elements which are either before or beyond the distance accommodated to are seen less clearly. Bearing these criteria in mind Strzemiński examines Renoir's Grand Boulevards (Fig. 4.17) and identifies:

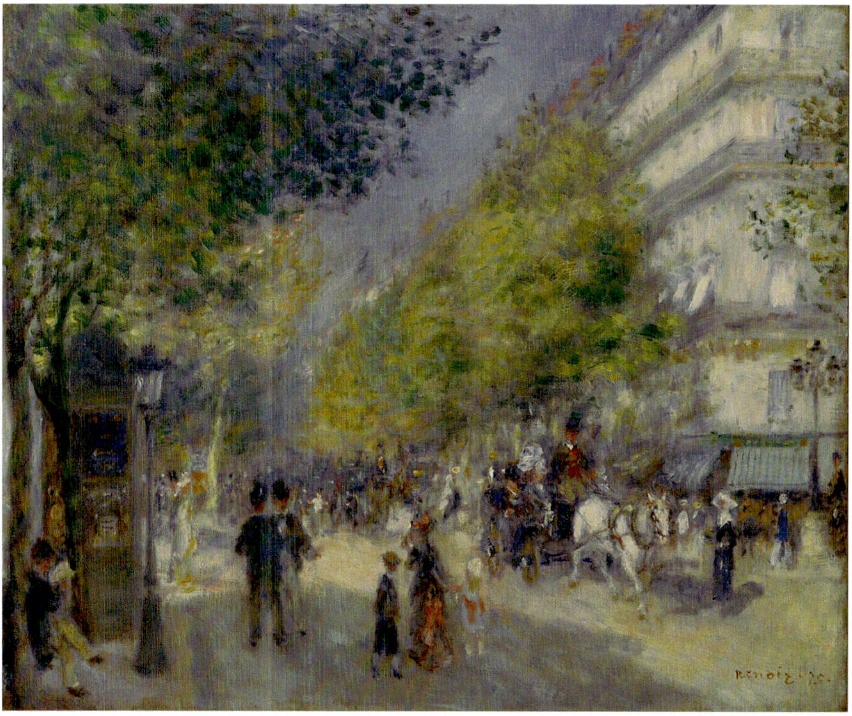

Fig. 4.17 Renoir's "Grand Boulevards" present several points of accentuated lucidity, identified by Strzemiński (Courtesy: Philadelphia Museum of Art)

i. Nearby and at about the same distance from the eye, two top-hatted men and a woman with children. These figures are all at the accommodation distance and all, with the exception of the child on the extreme right, are represented clearly. The child which is outwith the central visual zone is shown blurred as are all the other elements of the picture which are further away.
ii. On the right side of the picture two distinct accommodation points are found, one above the other. The lower shows a window of the house and the striped awning clearly, but the crowd and the lamp post which are nearer to the viewer as blurred; the upper point shows branches of the tree growing in front of the house clearly, but the house is now blurred.
iii. The carriage will provide another instance of the effect. The clarity with which the equipage is presented declines away from the horse's head. The head is depicted very clearly, the body less so, the driver even less and the passengers all but invisible.

All these focal points were determined, according to Strzemiński by the physiological principles of seeing—the eye being attracted to the most visually pronounced elements in the scene, for whatever reason. It was, perhaps, drawn to the horse's head by its nodding. Contrasts of colour and of lines, shapes, etc. were factors in the case of other foci.

Such "physiological", visual invitations within a scene would not be welcomed by Zeuxis. He wishes to portray the bison so that its picture can be readily recognized. Inclusion in the picture of a record of the perceptual experiences of the artist, Zeuxis would argue, merely obfuscates. It is of no more consequence to him than are the tobacco smoke and haze to farmer McDonald admiring his cabbages.

Notes

1. See Farish (1822). for a Precis of Farish's Argument Presented to the Cambridge Philosophical Society, See Booker (1979).
2. See Booker (1979) *History of Engineering Drawing*, an outstanding book which traces development of the discipline in an illuminating manner, from its very roots. (It is noteworthy that the Kakadu crocodile (Chap. 2) dismisses the problem of three-dimensionality by concatenating orthogonal views). Massironi (2002) discusses the transition from pre-technological to technical drawing.
3. For comparison of Hudson's Test and Construction Task, see Deręgowski (1968). For use of the Kwengo Callipers, see Deręgowski and Bentley (1968). Hudson (1960) seminal study of differences of ability to interpret pictures among distinct sections of South African populations caused widespread interest. The test booklet used in these investigations has been used by several workers, occasionally in modified versions, in order to clarify Hudson's findings (see Serpell and Deręgowski 1980). Grossly and misguidingly modified pictures, for reasons which are not made public, from the booklet have been published in several introductory texts. Some casual observations of responses to pictures in non-pictorial cultures are examined in the ultimate chapter of this book.
4. The drawing by S. Noakowski, whose talent Bartel greatly admired, shows an interior of a palace room. Application of Bartel's restitution shows that it grossly violates the rules of perspective. Noakowski's (1953) portfolio presents further instances of his cavalier treatment of the rules of perspective.
5. See ten Doesschate and Kylstra (1956) and ten Doesschate (1964).
6. See Deręgowski et al. (1994).
7. Between the central vision in which convergent perspective is perceived and laterally displaced vision where divergent perspective rules, there is a zone in which lines receding from the observer are seen by him as neither convergent nor divergent, but parallel. This presents the artist with a third, perceptually justified, choice of the way to evoke pictorial depth, a manner encountered in Persian miniatures and Japanese art.
8. See Deręgowski and McGeorge (2008).
9. See Deręgowski (2011), Cooke (1969, p. 150) for a photograph in situ.
10. See Laming (1959).
11. Hardonk (1999) in his extensive analysis of decorative band patterns points out that such patterns can often be perceived in a number of mutually exclusive ways. A linear sequence of alternating black and white squares may, for example, be seen as such a sequence or as a sequence of black squares on a white background or a sequence of white squares on a black background. Since patterns forming bands can be complex a variety of ambiguities can be created, providing an arena on which the perceptual system can demonstrate its ability to generate hypotheses (Gregory 1963).
12. There was a wave of cross-cultural studies concerned with perception of symmetrical geometric patterns, especially such as are used to test intelligence. It included studies by Ord (1971), Serpell (1971), Jahoda (1976), Jahoda (1978). Regrettably no analogous extensive studies of rock art exist.

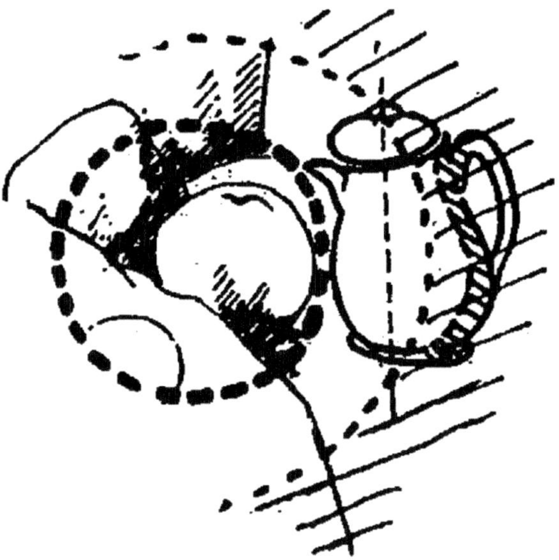

Fig. 4.N1 Cezanne's coffee pot as seen by Strzemiński (1974, p. 225)

13. See Darvas (2007) passim. The unqualified term *symmetry* is used in the present text in the manner of daily parlance and means symmetry about the vertical axis, such as the pairings OO, AA, and (). When other forms of symmetry are referred to they are appropriately qualified. For classification of symmetrical friezes, see Darvas (2007), Wasshburn and Crove (2004), Hardonk (1999).
14. See Deręgowski et al. (2000).
15. Marmor and Ravin (2009) present an excellent and richly illustrated review of effects of eye sight on creation and perception of works of art. It is a sequel to their earlier volume (Marmor and Ravin 1997).
16. See Marmor and Ravin (2009, pp. 22–23). Gombrich (1975) calls the erroneous explanation of El Greco's 'naïve belief'. See also Pirenne (1970, p. 43). Depiction by the artist of a distorted percept instead of the model would inevitably result in an infinite sequence of increasingly less and less satisfactory pictures. Model A perceived as A' would be portrayed as A' and therefore perceived as A" which in turn would be portrayed as A" and therefore perceived as A'". Looking at his picture of A, (A') the artist would see A". Would it not seem to him amiss?
17. See Pirenne (1970), whose experiments involving pinhole camera brought forcibly forth the issue remarked upon by Gombrich (1982), concerning the relationships between perspective psychology and geometry. The experiments also sustain earlier theoretical speculations by Strzemiński, which led him to postulate a visual pane normal to the line of sight as an explanation of Cezanne's style. For the relationship between Pirenne's data and Strzemiński's notion, see below.
18. Strzemiński (1958) A schematic representation of the effect postulated by Strzemiński is shown in Fig. 3.8. It shows projections of four vertical columns on a vertical plane normal to the line of sight. Columns a and d are at the same distance from the eye but the symmetrical projection of column a is narrower. Furthermore, projection of column d is asymmetrical, its central line is nearer to it left-hand edge than to its right-hand edge. The coffee pot on Cezanne's table (Fig. 4.14) shows analogous distortions. It does so because, according to Strzemiński the artist's central line of sight passed it by and rested close to the apple where the greatest variety of shapes and contrasts of light, colour and linearity are found. Strzemiński's figure (Fig. 4.N1) illustrates his conception.
19. Kubovy (1986, pp. 92–93).

20. An interpretation of the visual processes revealed by Cezanne's *Kitchen Table* was put forward by Loran (1943). The picture shows, it is argued, that the artist's eye level changed whilst he was painting. This is made apparent by the contrast between both coffee jug and tureen, and the large spherical pot. The former are shown by their about typical contours—the eye level cues therefore low, but the eye was at a much higher level when it was looking into the pot. (A tacit assumption is made that the tops of the three utensils were co-planar). Cezanne, Loran remarks, "had as habit of tipping vertical axes of nature to the right or left", as shown by depictions of the three utensils. Loran does not address the cause of this usage, nor does he consider gross asymmetry of the portrayals, an effect which dominates Strzemiński's argument. (This argument assumes, probably correctly, that the utensils were basically solids of revolution). In all three depictions the top of the supposedly vertical axis is leaning as if pulled along by the artist's gaze aimed at the particular focal point.
The emergence of these mannerisms in Cezanne's painting is not, in Loran's view, a result of profound intellectual reflexion. Cezanne was not aware of the perceptual vectors guiding his hand. The picture is a work of intuition, "accidental distortions arose from inner necessities of the particular problem at hand." This insightful opinion is probably correct. Awareness of the modus operandi of the perceptual mechanism is not a prerequisite of its use. An interesting but not widely explored artistic mannerism is that based on binocular disparity. Accommodation affects binocular disparity and leads to perception of "double images", the observer seeing as double targets on which he is not focusing and which are at a different distance from him than that focused on the target. Meinel (1973) thinks that such percepts could be explored artistically and brings to notice work of Evan Walters, a Welsh artist, who devoted much time to their study.

References

Bartel K (1958) Perspektywa Malarska. Państwowe Wydawnictwo Naukowe, Warszawa

Booker PJ (1979) A history of engineering drawing. Northgate Publishing Co., London

Cooke, C.K. (1969) Rock art of South Africa. Books of Africa, Cape Town.

Darvas G (2007) Symmetry: cultural-historical and ontological aspects of science-arts relations: the natural and man-made world in an interdisciplinary approach. Birkheuser, Basel

Deręgowski J.B. (1968) Difficulties in pictorial perception in Africa. British Journal of Psychology, 59; 195–204

Deręgowski JB (2011) On the perceptual significance of detached horns in Bushmen's depictions of animals; a note. Rock Art Research 28:271-272

Deręgowski J.B. (2014) Perspectival reversals of a solid: a phenomenal description. Perception, 43, 261–169

Deręgowski JB, Bentley AM (1986) Perception of pictorial space by Bushmen. International Journal of Psychology 21; 743–52

Deręgowski, J.B. & McGeorge (2008) Perception and the art of depiction of cylindrical objects. Perception, 37, 1879–1885

Deręgowski JB, Parker DM, Massironi (1994) Perception of spatial structure with oblique viewing: an explanation of Byzantine perspective. Perception 23: 5.13

Deręgowski JB, McGeorge P, Wynn V (2000) The role of left-right symmetry in the encodement of spatial orientation. British Journal of Psychology 91: 241–257

Ten Doesschate G (1964) Perspective, Fundamentals, Controversials, History. B De Graaf, Nieukoop

Ten Doesschate G, Kylstra J (1956) The perception of parallels. Ophtalmologica, 131: 61–65

Farish W (1822) On isometrical perspective. Transactions of the Cambridge Philosophical Society 1:1-19

Gombrich EH (1975) Mirror and map: theories of pictorial representation. Philosophical Transactions of the Royal Society of London 270:119–149
Gombrich EH (1982) The image and the eye: further studies in psychology of pictorial representation. Phaidon, Oxford
Gregory RL (1963) Distortion of visual space as inappropriate constancy scaling. Nature (London), 119: 678
Hardonk M (1999) Cross-cultural universals of aesthetic appreciation in decorative band patterns. Nijmegen Institute for Cognition and Information, Nijmegen
Hudson W (1960) Pictorial depth perception in sub-cultural groups in Africa. Journal of Social Psychology 52: 183–208
Jahoda G (1976) Reproduction of Kohs-type figures by Ghanaian children, orientation errors revisited. British Journal of Psychology 67:201–211
Jahoda G (1978) Cross-cultural study of factors influencing orientation error in the reproduction of Kohs-type figures. British Journal of Psychology 69: 45–57
Kubovy M (1986) The psychology of perspective and Renaissance art. Cambridge university press, Cambridge
Laming A (1959) Lascaux: Paintings and engravings. Penguin Books, Harmondsworth
Loran E (1943) Cezanne's composition: analysis of his form with diagnosis and photographs of his motifs. University of California Press, Berkeley
Marmor, M.F. & Ravin, J.C. (1997) The eye of the Artist Mosby, St. Louis
Marmor MF, Ravin JG (2009) Artist's eyes: Vision and the History of Art. Abrams, New York
Massironi, M. (2002) The Psychology of Graphic Images. Lawrence Earlbaum Associates Publishers, London.
Meinel E (1973) Peripheral vision and painting; a note on the work of Evan Walters (1894–1951) British Journal of Aesthetics 13: 287–297
Noakowski, S. (1953) Stanisław Noakowski; 20 reprodukcji, Wydawnistwo Sztuka, Warszawa
Ord IG (1971) Mental tests for pre-literates. Ginn, London.
Pirenne NH (1970) Optics, painting and photography. Cambridge University Press, Cambridge
Serpell R (1971) Discrimination of orientation by Zambian children. Journal of Comparative and Physiological Psychology 75: 312–316
Serpell R, Deręgowski JB (1980) The skill of pictorial perception: an interpretation of cross-cultural evidence. International Journal of Psychology 15: 145–180
Strzemiński W (1974) Teoria Widzenia. Wydownictwo Literackie, Kraków
Washburn DK, Cove DW (2004) Symmetry comes of age: The role of pattern in culture. University of Washington Press, Seattle

Chapter 5
On Cultural Press; If We Draw, What and How Should We Draw?

The skill of picture-making embraces the skill of picture-seeing,[1] but the two skills are distinct. The seeds of the skill of seeing pictures are universal, but their growth is shaped by the environmental and cultural press, which ensures that its proclivities are treated selectively.

There are no equally apparent perceptual seeds for the skill of drawing. It seems to be fostered by the drive to imitate; a desire to make something that evokes a perceptual response similar to that evoked by the model. Visual effects of such a drive may take many forms; an animal may be imitated by postures and movements of a Bushman dancer, it may be carved or moulded, and it can be drawn. Once the drawing is recognised a relationship between the two skills is established.

One can be an ardent viewer of pictures without being able to make pictures. There is a lopsided relationship between the two skills in which the skill of drawing is clearly subsidiary as it is culturally developed whereas perceiving involves an inherent biological process. Furthermore, depicting is not a skill essential for survival, not even for social cohesion. A pictureless culture can happily thrive. A traveller among the Ainu in the nineteenth century was surprised by the absence of pictures and the Ainu were fascinated by his ability to sketch. This seems to be the only report of members of a pictureless culture watching a draughtsman at work. Unfortunately, it is written, as travellers' tales are, and lacks the kind of detail that this author would appreciate.[2] Fortes[3] investigated how Tallensi children, brought up in the culture of northern Ghana, in which pictorial art was unknown, take up drawing. They did after some preliminary, apparently random scribbles, gain sufficient confidence with the use of pencils do depict some familiar objects, as shown in Fig. 1.9a. Thus demonstrating that the basic skill is easily accessible, but when it has no cultural worth it is unlikely to be developed.

It is obvious that an ability to see pictures is a prerequisite of the ability to make pictures. Are there pictureless cultures in which the skill of picture-seeing is not called for and therefore their denizens when presented with pictures may find them puzzling and opt for one of the polar choices, either dismissing them as meaningless patterns

J. B. Deręgowski, *On Picture Making and Picture Seeing*,
Vision, Illusion and Perception 4, https://doi.org/10.1007/978-3-031-23348-7_5

or treating the depictions as real objects—*trompe l'oeil* writ large. An experienced picture viewer does not yield to such notions.

Responses of the second kind to pictures by people living in non-pictorial cultures have been described by several authors. Theirs are startling, yet reliable, descriptions of exuberant reactions showing that pictures were misperceived as depicted objects. Such is the report of a great Scottish physician, missionary and traveller in Central Africa, Doctor David Livingstone, a guest of Shinte, an important chief in 1854:

"Shinte was most anxious to see the pictures of the magic lantern (an epidiascope) ... When I did go for the purpose, he had his principal men and the same crowd of court beauties near him as at the reception. The first picture exhibited was Abraham about to slaughter his son Isaac; it was shown as large as life, and the uplifted knife was in the act of striking the lad; the Balonda men remarked that the picture was much more like a god than the things of wood or clay they worshipped. I explained, the ladies listened with silent awe; but, when I moved the slide, the uplifted dagger moving towards them they thought it was to be sheathed in their bodies instead of Isaac's. 'Mother! Mother!' all shouted at once, and off they rushed helter-skelter, tumbling pell-mell over each other, and over the little idol-huts and tobacco-bushes: we could not get one of them back again. Shinte, however, sat bravely through the whole, and afterwards examined the instrument.[4]"

A strikingly similar occurrence was reported from Uganda by another witness fifty years later:

"When all the people were quietly seated the first picture flashed on the sheet was that of an elephant. The wildest excitement immediately prevailed, many of the people jumping up and shouting, fearing the beast must be alive, while those nearest to the sheet sprang up and fled. The chief himself crept stealthily forward and peeped behind the sheet to see if the animal had a body, and when he discovered that the animal's body was only the thickness of the sheet, a great roar broke the stillness of the night[5]"

These reports do testify to great independence of mind, courage and sagacity of both chiefs and also show, to a student of visual perception, that when extraneous cues are weakened by darkness when the rich background of the improvised screen, the buildings, the trees ... cannot be seen, the cues presented by the picture dominate the extent to which depicted objects are likely to be seen as objects. Both the Shinte court and Ugandan spectators saw the portrayals as real objects, not as pictures. The reports contradict any suggestion that pictorial perception is inherently distinct from perception of three-dimensional environment.

More telling evidence that pictures are correctly perceived even when presented as pictures in full daylight is that of Elmslie (1901), who wrote: "I have shown photographs of deceased persons known to them (the Ngoni), and they invariably turned away, some even running away in fear."

There is, however, also a body of data purporting to show that some cultural groups, are unable to perceive pictures. Observations by some anthropologists suggest that this may be the case.

"I have had an experience of this kind, similar to that reported from many parts of the world by those who have had occasion to show photographs to persons who have never seen a photograph before ... In the instance to which I refer, a Bush Negro woman turned a photograph of her own son this way and that, in an attempt to make sense out of the shadings of greys on the piece of paper she held. It was only when the details of the photograph were pointed out to her that she was able to perceive the subject.[6]

Another observation by a missionary comes from Malawi (then Nyasaland):

"Take a picture in black and white, and the natives cannot see it. You may tell the natives: 'This is a picture of an ox and a dog'. Well, perhaps they will tell you what they think this time. If there are boys about, you say: 'This is really a picture of an ox and a dog. Look at the horn of the ox, and there is his tail!' And the boy will say: 'Oh! Yes, and there is the dog's nose and eyes and ears!' Then the old people will look again and clap their hands and say, "Oh! Yes, it is a dog!".[7]

Both pieces of evidence have stings in their tails as regards inherent inability to see pictures, for they both show that this difficulty is easily banished by drawing the viewer's attention to some distinct feature of the portrayal. Visual association of that feature with an object is their key to the world of pictorial perception. Similarly, a responder to a Rorschach inkblot may maintain that it is only an inkblot until asked what a particular contortion of the inkblot's edge looks like. The Mekan of Ethiopia[8] saw the surface patterns printed on paper as of no interest at all. The paper, to them an unknown substance, absorbed all their attention; they fingered it, sniffed it, peeled little pieces and chewed them. When on another occasion pictures printed on coarse cloth were shown to them their interest in the background surface vanished. Even so, Mrs Muldrow, the experimenter, had on occasion to foster recognition by drawing attention to a feature of the depiction, just as it was done in Malawi a hundred years earlier. She had to urge particular use of the perceptual cues, just as a skilled microscopist does a tyro microscopist; a skilled industrial inspector does a tyro industrial inspector, and a safari guide does a tourist. The result of such instruction is to subdivide the swarm of stimuli assailing the perceptual system and draw attention to specific cues. In the case of pictorial perception this means recognition of the crucial pictorial elements and relegating the stimuli such as those associated with the material of the picture's surface into secondary awareness.

Thus, the degree of experience of pictures defines the extent of the viewer's secondary awareness. A person familiar with pictures would treat paper and coarse cloth as the base material for the production of the picture and so consign them to secondary awareness. For the Mekan, who were unfamiliar with pictorial depictions, paper, a novel substance merited full attention and the coarse cloth was worthy only of their secondary attention. This effect of selectivity of attention and therefore of the secondary awareness is responsible for the difference between responses of some children to different tasks involving the same pictures. A child may be equally effective at naming a picture of an object (say, a toy) as the object itself, and yet attain significantly higher scores on a sorting task when objects rather than pictures are used as tokens.[9] Pictures do not cause difficulties because depictions on them are not *seen, but* because they share the attribute of flatness of the surface, whilst solid objects do not, and may also have pronounced individual visual characteristics.

The effect is not to be observed only in the course of psychological tests and juvenile parlour games. When Scottish women were asked to memorise the locations on a table of a set of household objects or the locations of their pictures, they recalled the locations of objects better than they did locations of pictures.[10]

Early Nyasaland and recent Ethiopian reports show that the notion that a graphic design represents a three-dimensional object is easily introduced to observers from non-pictorial cultures by directing their attention to some salient feature of the representation; a dog's snout, a cow's horns. Once this is grasped the observer builds a sensible percept and recognises other parts of the design as related to the nuclear element. This process of searching for a meaningful entity is similar to the skill routinely exercised in any visual search. It is used when hunters search for a game, when the fairy searches for a needle in a haystack, and also when one searches for a familiar face in the crowd of arrivals at a railway station. Here, however, the search gains special significance because it is done in pictorial space and establishes that a picture represents an object.

Recognition that a picture is a representation does not ensure errorless perception of pictures. A depicted tortoise can still be mistaken for a snake emerging from a hole, as the Flying Doctor Service report shows. Much depends on the manner of representation.

The Dowayos of North Cameroon could not distinguish between a picture of a leopard and that of a lion, presented on postcards, but they thought outline drawings of human beings "a major wonder" and came to the anthropologist specially to see them.[11] Their readiness to recognise the peripheral contour would please Zeuxis Primus greatly.

To dispel the impression that mis-perceptions of pictures (from the artist's point of view) occurs only in remote populations, which these pieces of evidence might suggest, let us quote from correspondence of Jean Baptiste Corot, the French landscape painter, with his agent in Paris. Corot describes himself at work watched by Simon, a local rustic:

"... one paints, one paints! Oh! The beautiful chestnut-coloured cow, standing up in the marshy grass. I will paint it ... Crack! There it is – splendid, splendid; goodness, how striking! Let us see what this peasant will say, who stands watching me paint but is too shy to approach. 'Hi there Simon!' Good! Here is Simon approaching and looking.

'Well, Simon, what do you think of it?'

'Oh, really, sir, it is very beautiful.'

'And do you see what I wished to represent?'

'I think I see what it is; it is a large yellow rock you have placed there!'"[12]

Since perception of pictures involves the same perceptual mechanism as does perception in general there appears to be no ground for regarding the basic ability to perceive pictures as variable within the human race. Cultures differ in skills which they cultivate and cherish. One should be wary of notions that a picture is a perceptual entity grossly different from the surrounding objects, that when viewing it attention needs to be distributed differently. Cross-cultural differences of interpretation may occur, but such differences are also found within cultures; artists and art critics often talk about pictures in a manner incomprehensible to laymen.

Depiction, conveying of a visual impression onto a surface, is difficult because the viewer of a model sees a visually rich percept derived from a plethora of stimuli, but reproduction of all these stimuli is not possible. A selection has to be made, just as it is made daily when looking for an object, but the root of the skill of depiction is not equally apparent. Once the first fruits of the skill—pictures—have appeared, its growth could progress by mere imitation and elaboration, but its very origin is obscure. One can, however, with reasonable conviction, define various constituents of the skill beginning with recognition of the model as an entity distinct from the background. What must follow is the detection of those features of an object which isolate it from the background. These are likely to be peripheral contours. Some peripheral contours look much more salient than others and foster recognition of an object—these contours are therefore called typical.

The easiest objects to draw are those that proffer a single planar typical contour which can be copied onto a parallel, in reality or notionally, surface of a rock face, paper or canvas. A playing card or a cardboard cut-out is easy to depict, but so are other objects which, although they are more assertively three-dimensional, have their typical contours in planes parallel to the artist's surface. The bisons of Altamira were, obviously, close approximations to such objects, if one can imagine the artist watching a group of bison walking steadily by. However, many objects deviate from this ideal of a single planar contour. the artist has to contend with these deviations.

If perceptually distinct parts of an object have planar contours in planes at angles to each other, the artist may translate such contours into a single plane, as the Ancient Egyptians did when depicting the Holy Cow (Fig. 1.18b) and the Australian artist did when portraying crocodiles (Fig. 1.16). The elements are then represented at their best by their typical contours. It is only the implied relationships between them that some would question.

When a model presents several contours in different planes which are linked, as are the faces of a cube, each sharing a contour line with a neighbouring face whose contour is in a different plane, the difficulty of drawing is greatly increased, as the drawings obtained by Bartel from illiterate agricultural workers show (Fig. 1.19).

Nor does literacy ensure smooth sailing in the world of depiction. A geometrically very simple solid, a clegg,[13] is shown in Fig. 5.1. It is defined by an orthogonal intersection of two co-planar cylinders of equal diameter. It has two elliptical typical contours in planes intersecting at right angles. No view of the figure presents both elliptical contours in their true shape; no shadow projected by Mr Bursill can do so, although the projection changes dramatically when the figure is rotated. The figure can project a circle (to which each contour contributes an even share); a square (wherein both contours are as straight lines forming diagonals and the sides are peripheral contours) and as an ellipse (wherein one of the contours is shown in its true shape, but the other is reduced to a short straight segment (Fig. 5.2).

Undergraduate students handed the clegg and told to examine it both visually and by palpation for three minutes, and then to draw it from memory, found the task extremely irksome. The clegg's two typical, and perceptually equally weighty contours are mutually orthogonal and intersect, not just abut, as elements of the Kakadu crocodile do. Thus bound the contours cannot be cast on a single plane,

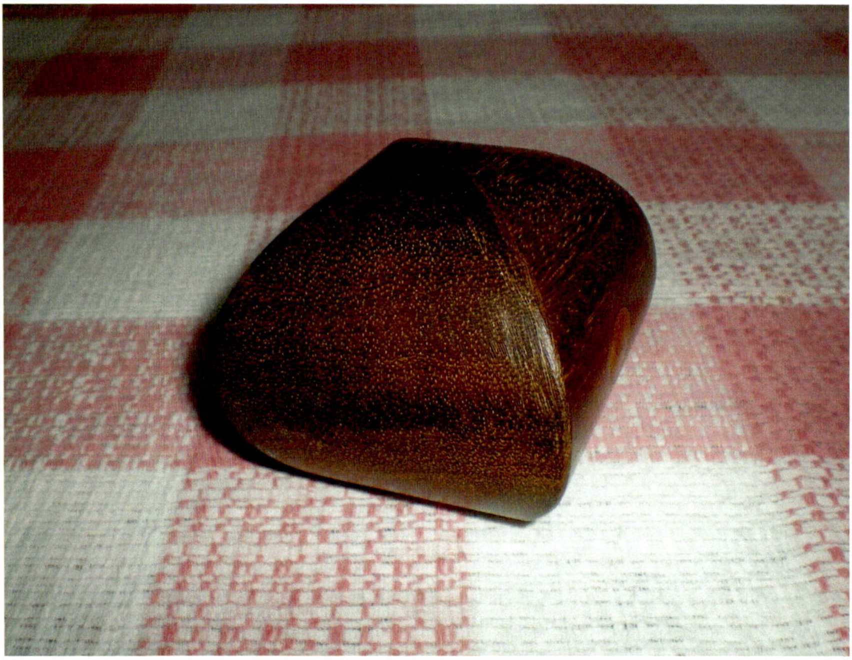

Fig. 5.1 A clegg[13]

unchanged in the manner of the typical contours of the body parts of the crocodile, which are merely concatenated, not integrated. None of the random group of ten students achieved a convincing drawing. It is not the difficulty that Zeuxis and other draughtsmen from nature experienced but the experiment confirms the importance of typical contours (There is a rank order, it seems, of typical contour entanglements with increasing difficulties of drawing represented by: (1) Kakadu crocodile; (2) a cube; (3) a clegg).

Typical contours can be bent double without loss of shape, thus creating, for example, a contour of an animal looking towards its rear. The shapes of the two elements are not altered by this, and are clearly co-planar in the surface of the picture. Subtler change, change of the shape of the typical contour, evokes pictorial depth. The picture of the two Bushman hunters and a dog discussed earlier (Chap. 2) provides an instance of this. Roger Fry describes the taller hunter as: "… the silhouette of a most complicated gesture with foreshortening of one thigh and crossing of the arm holding the bow over the torso" (Fig 2.10). The lines formed by the hunter's legs and his body form an angle that suggests depth (Fig. 5.3). Pictorial depth is also evoked by the rhebok facing away from the viewer, in another Bushman picture (Fig. 5.4a). The central lines of the ears, the horns and the neck converge just as do lines of perspectival drawings which converge to a point on the horizon—the line most distant from the viewer. Central lines which are about normal to the horizon (or notional horizon) are clearly those of elements nearly extending from the viewer

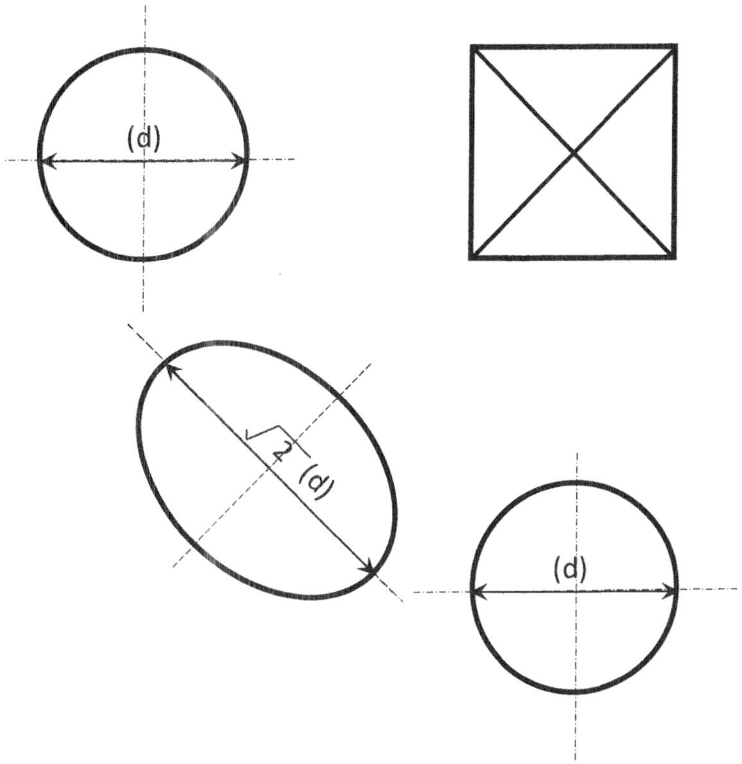

Fig. 5.2 The attributes of the clegg and some of its views. Two identical coplanar, mutually orthogonal, cylinders of diameter d intersect to form a clegg. Viewed normally to the plane of the cylinders' axes, the clegg casts a square of side d. Viewed along the axis of one of the cylinders it casts a circle of diameter d. Viewed along the bisector of the angle between the axes of cylinders and with the eye coplanar with the axes it casts an ellipse (axes, $d\sqrt{2}$ and d). Two such ellipses form the two visible ridges when the square peripheral contour is seen

and therefore of particular significance (Fig. 5.4b). Both Bushman pictures embody the same devices as those used in the Gossamer Girl (Fig. 2.4).

The artist who created the hunters lived free of the influence of Ancient Greeks, as did the Mayan artists who decorated the walls of Bonampak sometime in the eighth century A.D. which show use of foreshortening. These instances and the analysis of Abbé Breuil's tracings of Altamira bovids (Chap. 2) show that artists free of Ancient Greek influence were able to distort typical contours to create foreshortening.

A particularly telling use of foreshortening is present in Li Di's 12th–13th century picture showing herd boys and buffaloes in a downpour of rain (Fig. 5.5). The herd boys, dressed in cloaks of reed and hats of bamboo are on top of the buffaloes. The buffaloes are of particular interest. The one on the right presents the typical contour of the species in the manner which Zeuxis Primus would find admirable. The buffalo on the left has its head turned to the viewer, its spine is twisted and a segment of its

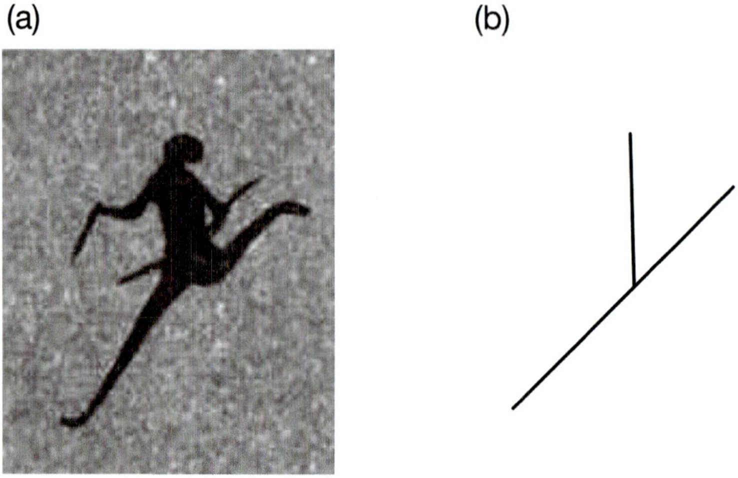

Fig. 5.3 **a** A running hunter. An element of the Bushman hunt (Fig. 2.10) (From: Tongue 1909). **b** Central lines of the hunter

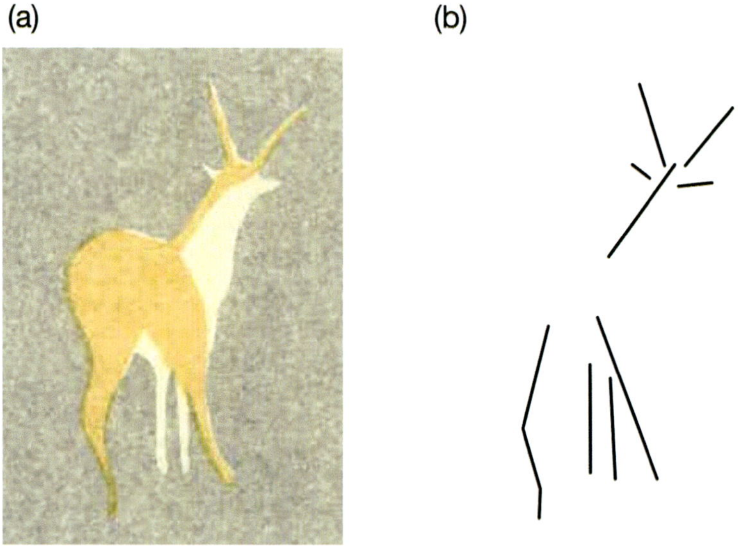

Fig. 5.4 **a** A rhebok as seen from behind (From: Tongue 1909). **b** Central lines of the rhebok horns and ears converge to a point on the central line of the neck and thus define the most distant point of the putative three-dimensional array (point Z in Bartelian notation (Chap. 3) (The rhebok's rotund hindquarters evade representation by a central line, as do those of a digging woman shown in Fig. 1.9b)

(a)

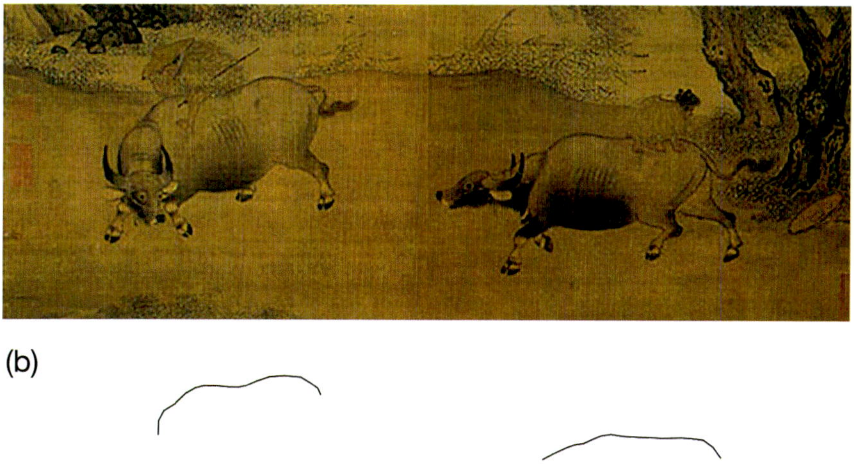

(b)

Fig. 5.5 **a** A game in the rain, enjoyed more, one suspects, by the herd boys than by the buffaloes (For information about the role of the buffaloes in the culture of the day, see Bo Liu 2014).[20] **b** Shows the contours of the two animals

typical contour has been replaced by a peripheral contour and remains in the picture as a significant contour. The animal is in violent motion. It is not, probably, finding the game as enjoyable as does the herdboy on its back.

Indeed, Bushman artists could distort the contour in even subtler manner by altering it to record, not changes of spatial orientation, but changes with the passage of time, as when a man, by degrees, turns into a frog (Fig. 1.24). Isolated instances of foreshortening are to be found in other ancient cultures, in which they failed to flourish.[14] Natural perceptual impulses were suppressed by the prevailing cultural ethos, not because they created pictorial space, but, one suspects, because foreshortened portrayals are more difficult to recognise than portrayals showing undistorted typical contours. It was a series of foreshortenings that converted the man into the frog; the intermediate portrayals of the series are clearly ambiguous, they are more difficult to draw than the initial and final pictures, they are also in isolation difficult to recognise as transitory forms.

Dominance of intact, unambiguous typical contours is found among people with little pictorial experience. When two pictures of elephants (Fig. 5.6), one portraying it from above and therefore foreshortened, and another splay-legged, were shown to illiterate labourers, who were asked which one they preferred, there was an overwhelming preference for the complete, splay-legged beast (with the exception of one man, who did not like it, for it showed an elephant "wildly jumping about", hence dangerous).[15]

The reason for the flourishing of foreshortening in only some cultures cannot be sought in inherent differences of perceptual *capacities* (as distinct from *skills*). It lies in the social realm. The replacement of the two-eyed profile, the *de rigueur* portrayals of men by 19th century children, by frontal views of the face, shows that such forces are present even in nurseries.[16]

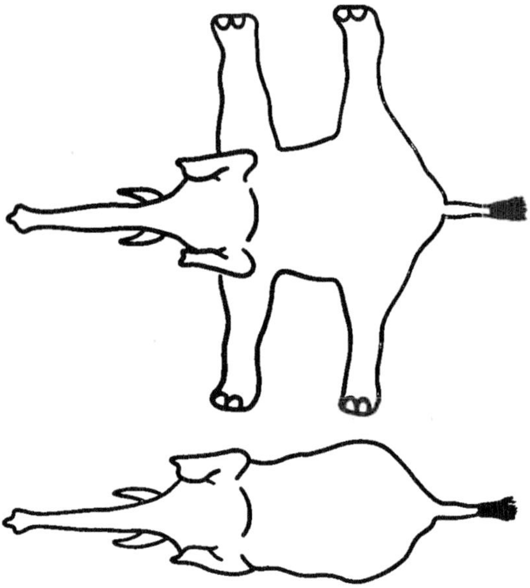

Fig. 5.6 How to draw a recognisable elephant?

The perceptual tesserae which the artist uses to construct his picture may originate from experiences involving either central vision or non-foveal vision. They may originate from views beheld by the eye in differing stances. Their choice and their arrangement is affected by the cultural usage of the day. This may lead to the creation of patterns, whose beauty overshadows veracity, as in the miniature from the Aberdeen Bestiary. In it (Fig. 5.7) only the three skeps are unmistakable. Individual bees do not look like bees in flight, nor are bees ever seen flying in such rigid formations, but the skeps are there and bees fly into skeps, therefore, with minimal induction, the "bees" are bees.

The eye is avid to make sense of the visual environment and, guided by its present perceptual set and previous experience, it marshals sensory stimuli into significant entities (Had the Mekan not been so taken with the strange new stimulus, paper, they would have seen pictures on it). The problem of marshalling stimuli within the eye in order to recognise the existence of such a new entity as a picture is not essentially different from that of marshalling them to recognise a particular face or a particular bacterium. All people including those living in pictureless cultures (if there are still such) are as capable of seeing pictures as an European urbanite is of spotting a zebra grazing in the bush.

The making of perceptual sense calls for adjustments of the impact of optically distinct features so as to derive a cohesive percept. This implies tolerance of discrepancies. Tolerance, for example, of discrepancy between the rate of convergence of

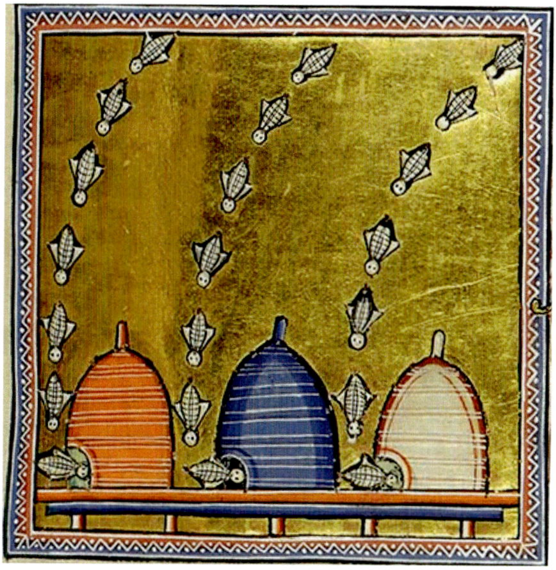

Fig. 5.7 Homing bees in a thirteenth century miniature (Aberdeen Bestiary)

receding lines and of depth-compression in pictures done in perspective. Such tolerance is, as shown in the third chapter, readily exercised. There are cases, however, when its limits are breached. Then impossible figures, such as the two-pronged trident (Fig. 1.10) are seen.

Most pictures are apprehended instantly just as are most objects, but there are pictures, as there are objects, which are difficult to apprehend. Some pictures present visual double entendres, such as Fig. 5.8, the correct perception of which (that is, perception intended by the artist) rests, in part, on information other than that contained in the visual stimulus. The nature of this 'external' information may differ with viewers. Thus the figure which is graven on a rock in South Australia is said by some to show the head of a salt water crocodile, a tenet that seems to some, including the present author, who has never seen a salt water crocodile, entirely plausible. However, others including some local people say that it is a picture of yariba, a magical device, which most of the readers of this book and its author have neither seen nor are likely to see. There is ecological evidence against the first interpretation—salt water crocodiles were never found in South Australia. Not a crocodile, then. Is it likely to be a yariba? The learned disagree.[17] The learned have also, and much more volubly, disagreed about a small rock painting in Brandberg, Namibia. Is it a portrait of an immigrant white lady of high birth, or of a local youth on his way to an initiation ceremony? (Breuil 1955), Lewis-Williams and Dowson, 1999)

Such pictures would not meet with Zeuxis's approval. How can something be called a picture when one is not sure what it is a picture of? To make a proper picture Zeuxis must choose from remembered real world shapes such as those of bison, shapes which drawn on the rock face will retain its meaning. Enduring salience is the

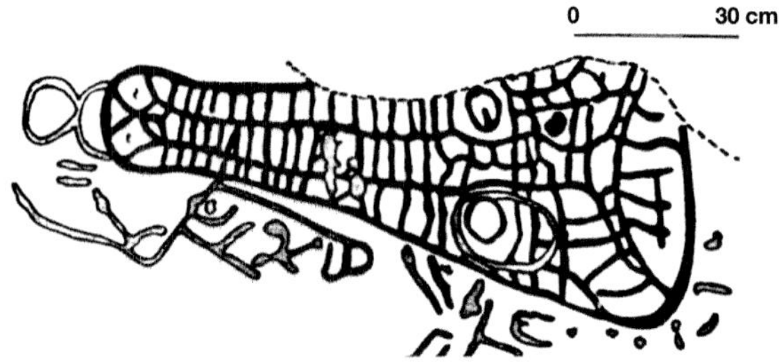

Fig. 5.8 A yariba or a crocodile? An Australian problem

criterion. Typical contour, because of its power of evocation, is the obvious choice. One must beware of an aesthetic impulse.

Zeuxis Primus chooses the typical contour as the key, convinced that those painters that will come after him would do the same.[18] They will do so, not only because they see the same features as salient, but also because Zeuxis has let out the genie showing that a picture can be made and inevitably suggesting how it should be made. But, being artists, they will not slavishly submit and obediently follow the narrow path of invariable typical contours, for there are, as Bushmen have shown, great gains to be obtained by deviating from it. How else could one record the amazing transmogrification of a man into a frog (Fig. 1.24); or the joy of riding a buffalo in the rain (Fig. 5.5)?

The striking Isfahani miniature (Fig. 5.9) carefully signed and dated (1650) by its author, shows no pronounced typical contours—it does not meet Zeuxis's criterion. An Italian colleague thinks that the pale yellow twisted mass represents an uncooked tagliatelle. Encouraged to press on; the viewer cautiously advances, in the manner of the Nyasaland boy. A suggestion that it is a portrait of a sleeping dervish is readily accepted. A curled-up dervish tired by whirling in an ecstatic dance, perhaps, is taking a nap in a quiet corner. The shape of his body is distorted, just like that of the bodies of elands (Fig. 1.21b) painted by a Bushman artist. The same perceptual mechanism generates similar artistic devices in distinct cultures.

In the attempt to show ubiquitousness of perceptual processes and their impact on picture-making, evidence from several distinct cultures was invoked. The concern was with depictions rather than with cultures, although *intra-cultural* variations in picture making may contribute significantly to understanding of the processes. Cultures with notably rich tradition of picture making seem particularly fit for the purpose. The culture of south African Bushmen whose rock art has been repeatedly referred to above is such a culture. Another culture in which picture making was highly esteemed and which can boast of a body of well-preserved ancient art is that of China; a glance on a segment of an ancient scroll is sufficient to show its vigour. It is the Qingming Scroll, a picture created in the 12th century, when the Song Dynasty ruled, by Zhang Zeduan.[19]

The vigour of picture making in a culture can be measured by the extent to which its art complies with Zeuxis Primus's basic desideratum that the depicted model should be readily recognisable; further, the ease with which the depiction is a

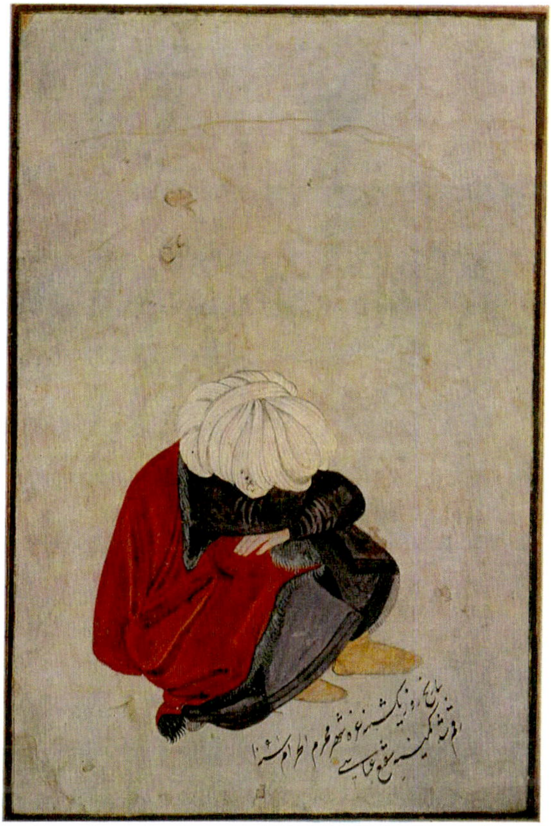

Fig. 5.9 A dozing dervish. An Isfahani miniature by Shafi-i-Abbasi (1650)

comprehensible modification of the percept, as demonstrated by the Bushman record of transmogrification of a man into a frog (Fig. 1.24) and, even further at a remote extreme, by embodiment of the sense of humour, as freely displayed by the Bushman lion of Tyffelfontein (Fig. 1.14), which shows how free the artist felt.

On the Qingming sroll (Fig. 5.10) a cow has broken loose and runs down a slope dragging its broken stay, in a most uncowlike manner. Three men waving their arms, and probably shouting, run after her and cause her to run even faster. Two palanquin bearers trot behind them, and behind the palanquin there stretches up the hillside, like a lizard's tail, a train of men all bound for the town.

Another cow impassively ruminating stands in the shadow of a roadside tree and watches the chase, as does the equally impassive, portly codger seated outside a house further down the way. A testimony to the artist's sense of humour.

Yet another cow reposing in the shade next to the impassive ruminator cannot even be bothered to look. The impassive ruminator is clearly foreshortened as are many animals depicted on the scroll.

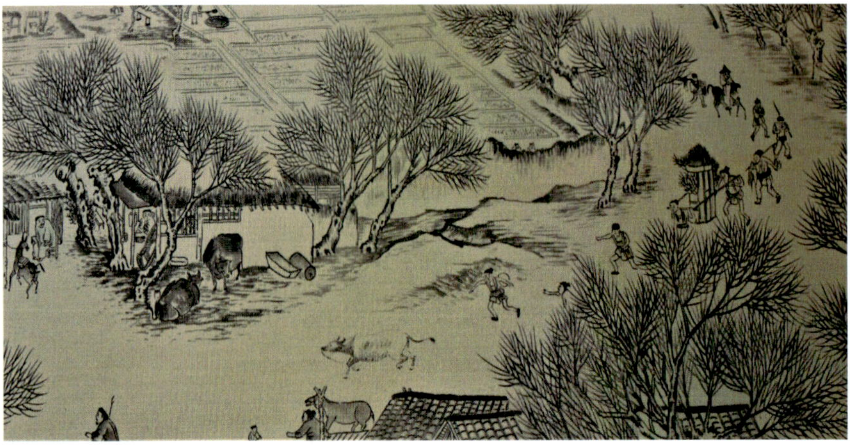

Fig. 5.10 The cow that got away. It is being impassively watched by another cow and by a codger seated outside the entrance to a house

Startlingly effective use of foreshortening enlivens Li Di's twelfth-thirteenth century hanging scroll and supports the argument advanced earlier, against the cultural monopoly of the usage of foreshortening.

On the Qingming scroll, in downtown tea-houses rectangular tables await customers. The tables are drawn in the Byzantine reverse perspective. This perspective is not, however, dominant. It is present and easily apprehensible in representations of smallish rectangular pieces of furniture, such as tables, benches, stalls, less detectable in depictions of larger surfaces, such as short roofs and absent in depictions of larger rectangular surfaces, such as the long roof of the tea house in Fig. 5.11 which has a hint of convergent perspective, as do the junks moored nearby. The artist was not, it seems, rigidly following any convention. This does not show lack of interest in perspective, rather it shows an inclination to explore it without firm allegiance to any particular notion.

The significance of universality of perceptual attributes was already discussed, but these studies make yet another point.

A 19th century adage 'the eye is like a camera' quoted with assurance by the pedagogues wielding tawses was by the end of that century threatened by the studies of illusions. 'There are more trees in the forest than can be seen from a single stance.'

The different conceptual stances from which a picture can be viewed can be determined by non-visual associations of the culture. An acquaintance of mine viewing a picture of a three-legged stool, standing on two legs, leaning against the wall, the third leg pointing upwards, said 'a broken stool.' 'He has not read Freud', whispered my colleague. Maybe not, a cultural chasm was revealed, but it did not question the correctness of his perception.

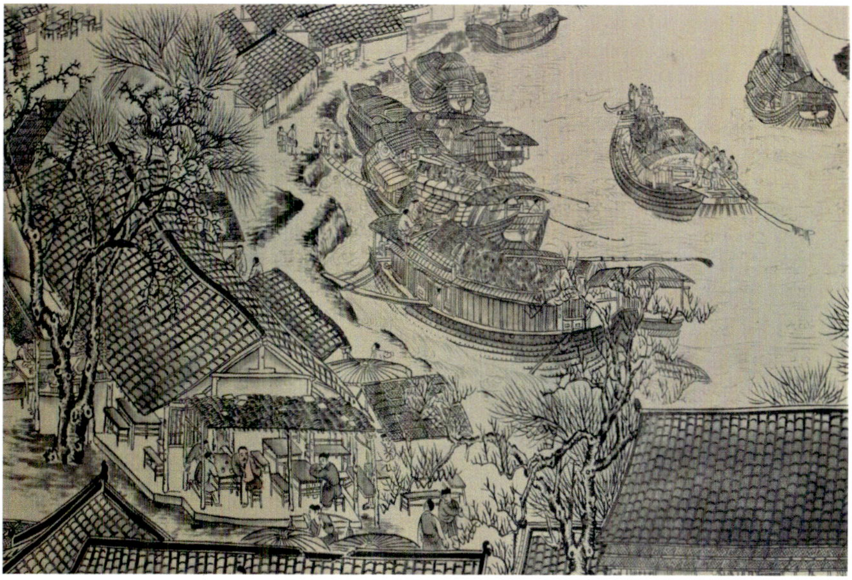

Fig. 5.11 Junks moored by a tea-house. In the well patronised tea-house all table tops are drawn in Byzantine perspective, the heavy tiled roof, in contrast, grows narrower with perceived distance

Notes

1. For general discussion of skills, see Welford (1968). For discussion of the skill of pictorial perception, see R. Serpell and J.B. Deręgowski (1980).
2. A. H. S. Landor (1883).
3. M. Fortes (1940), and (1981)
4. D. Livingstone (1857) p. 298.
5. A. B. Lloyd (1904).
6. M. J. Herskovits (1948), also in (1959) and quoted in M. H. Segall, D.T. Campbell and M. J. Herskovits (1966).
7. H. P. Beach (1901).
8. J. B. Deręgowski, E. S. Muldrow and W. F. Muldrow (1972).
9. I. E. Sigel and R. R. Cocking (1977). Both authors are noted for their contribution to psychological studies of children.

   See also J. B. Deręgowski and R. Serpell (1971).

   Z. S. Klapper and H. C. Birch (1969) have shown that imitative movements evoked by hand tools are more vigorous than movements evoked by pictures of these tools.
10. J. B. Deręgowski and G. Jahoda (1975).
11. N. Barley (1986).
12. Corot's letter quoted by D. C. Thomson (1902).
13. So called after John Clegg, Australian archaeologist and man of many talents, who brought this solid to the author's attention. Mr. Bursill, of the first chapter, would be quick to point out that characteristic shadows cast by a clegg are circles, ellipses and squares. So unlike the characteristic shadows of the cube, its distant cousin, which are only squares.
14. E.H. Gombrich (1962) pp. 122–123.
15. W. Hudson (1962).

16. Two children's drawings, one showing a two-eyed profile Sully (1895) and another done recently by a girl of seven years showing a frontal view of a face, are shown in Fig. 1.28. Wilson and Wilson (1982) analysed the gradual change of stylistic preference of juvenile artists between the penultimate decade of the nineteenth century (70% of drawings by Italian children show the two-eyed profile) and the middle of the twentieth century (no such drawings were recorded in an American study). The title of their article summarises their findings: "The case of the disappearing two-eyed profile: or how little children influence the drawings of little children."
17. R. G. Bednarik (2017) discusses this figure as well as other rock pictures. The notion that it is of a salt water crocodile was put forward by C. P. Mountford and R. Edwards (1962). R. M. Brent thought that it represents a yariba. Bednarik points out that there were never salt-water crocodiles in South Australia, and discusses the problem.
The list of influences that Zeuxis wishes to bar from his art, given earlier, omits the influence which affected his work but of which he was only partially aware, that of ease of execution. His choice of the surface on which to paint is so affected and he is conscious of this, but he is not conscious that the ease of execution of movements in the course of painting affects the result. In order that the reader may not leave this book in a state of equal innocence—this note. When copying vertical or horizontal lines, lesser errors are made than those made when copying oblique lines; and when copying oblique lines errors are those of drift towards those of the two principal orientations with which they make a smaller angle. (An oblique at 30° to the vertical drifts towards the vertical, an oblique at 30° to the horizontal drifts towards the horizontal.) Truthful depiction of oblique lines is therefore more difficult than depiction of vertical or horizontal lines, a fact acknowledged by Breuil and Obermaier (1935) by their comments about orientations of uplifted arms of mannikins in the Altamira Cave.
When Scottish children were presented with stick models of giraffes and asked to draw them they made the kind of errors adumbrated above. Such trends are also found in rock art and probably explain the "sitting giraffes" of the Sahara (J. B Deręgowski and F. Berger 1997).
18. When Zeuxis Primus drew his bison (Chap. 2) to make a picture that would beguile people to think, on seeing it, that they see a bison, he was not concerned with aesthetics.
Many millennia later Clive Bell (1924), an art critic, asserted that he had "little doubt that what makes the palaeolithic art so bad is a preoccupation with exact representation." (Bell 1924, p. 23). Bell thought that *Significant Form* is at the core of a work of art and is the trigger of unique aesthetic emotion. The nature of the significant form is not clearly described, but it is the form that is used as object of emotion. Bell repeatedly stresses the importance of aesthetic emotion. "To appreciate a work of art we need to bring with us nothing but a sense of form and colour and knowledge of three-dimensional space" (Bell 1915, p.27). Two of the elements listed by Bell have been considered in the present volume, but not the matter of colour. The relationships among the three are not elaborated by Bell, hence vagueness prevails. This vagueness and the lack of empirical tools (other than questionnaires) for attacking aesthetical notions, has greatly discouraged experimental psychologists from evaluating the theoretical significance of Bell's observations. All was quiet on that front until 2013, when Ishizu and Zeki armed with a magneto metric scanner (a device for observing brain activity) arrived, and conducted a series of experiments, one of which showed that judgements of a perceptual attribute (brightness) and affective attribute (beauty of pictures) engaged distinct brain systems. Another compared responses to beauty of music and paintings and found close correlation. The judgement of philosophers of these findings seems uncertain, bearing in mind great uncertainties of aesthetics (Dziemidok 1988). The response of experimental psychologists would probably be: "could we please have some more evidence and, if possible, fewer words." Related discussions of the problems involved are to be found in Zeki and Ishizu (2013) paper on Francis Bacon's art and its subversive impact on the normal neural representation. This probably arises because the two specialised systems of the brain (Ishizu and Zeki 2013) are at loggerheads, the perceptual system (guided by the typical contours?) identifies the face but affective (aesthetic) system finds this identification questionable.

Zeki's studies show that the elusive aesthetic factor should be taken into account in *perceptual* studies of works of art. This notion is also present in Weller's (1990) observation that reports of aesthetic experiences of art impulses so strong as to distract viewers from the humdrum daily sensations are not difficult to find in literature.

This of itself does not allow us to assume that the matter is worth of further scientific study, any more than frequent reports of ghost sighting persuades us to study ghosts. In a report on a recent investigation involving interviews and carefully devised questionnaires Csikszentmishalyi and Robinson (1990) concluded that "the aesthetic experience is one of the most ingenious vehicles for making life richer, more meaningful". However, its effect on pictures' recognizability remains obscure.

19. This masterpiece which is kept in the National Museum in Beijing was kindly brought to the author's attention by Prof. G. MacCormack. The scroll is about 10 inches (25 cm.) high and over 12 feet (372 cm.) long. The photographs in the present text are of a presentation facsimile of the scroll, owned by Prof. G. MacCormack. The photographs were taken by Mr. P. Bates.
20. The scroll is kept in the National Palace Museum, Taipei.

References

Barley N (1986) The innocent Anthropologist. Penguin Books, London

Beach HP (1901) Geography and Atlas of Protestant Missions. Student Volunteer Movement for Foreign Missions, New York

Bednarik RG (2017) Palaeoart of the Ice Age. Cambridge Scholars Publishing, Newcastle upon Tyne

Bell, C. (1915) Art., Chatto & Windus, London

Bo Liu (2014) The multivariant imagery of ox in Song painting. Journal of Song-Yuan Studies, 44, 33–84

Breuil H (1955) The white lady of Brandberg. The Tranion Press, London

Breuil H, Obermeier H (1935) The cave of Altamira at Santilana del Masr, Spain. Topografia de Archivos, Madrid

Caleti AV (1949) Bonampak la Cuidad de los Muros Pintados. Supplemento al T.III (1947–1948) de los Annales del Instituto Nacional de Antropologia e Historia. Mexico

Csikszentmihalyi, M. & Robinson R.E. (1990) The Art of Seeing: An Interpretation of the Aesthetic Encounter. The J. Paul Getty Museum, Los Angeles

Deręgowski JB (1968) Pictorial recognition in subjects from a relatively pictureless environment. African Social Research 5: 356–364

Deręgowski JB (1994) Kazimierz Bartel (1882–1941). Psychologie und Geschichte, 5: 246–260

Deręgowski JB, Bentley AM (1986) Seeing the impossible and building the likely. British Journal of Psychology 78: 91–97

Deręgowski JB & Berger F. (1997) "Sitting Giraffes" Sahara 9:87–92

Deręgowski, J.B. & Jahoda, G. (1975) Efficacy of objects, pictures and words in a simple learning task. International Journal of Psychology, 10 19–25

Deręgowski JB, Parker DM (1995) Viewing angle and the perceived orientation of pictorial elements: Geometric or representational effects? Perception 24: 1139–1153

Deręgowski, J.B. & Serpell, R. (1971) Performance on a sorting task: A cross-cultural experiment. International Journal of Psychology, 6, 273–281

Deręgowski JB, Tatler BW (2016) The eye of the illusion: Variation in perceptual phenomena within an inverted percept. Perception 46: 100–108

Deręgowski JB, Muldrow ES, Muldrow WF (1972) Pictorial recognition in as remote Ethiopian population. Perception 1: 417–425

Duncan HF, Goulay N, Hudson W (1973) A study of pictorial perception among Bantu and White primary school children in South Africa. Witwatersrand University Press, Johannesburg

Dziemidok, B. (1988) Controversy about the aesthetic nature of Art. British Journal of Aesthetics, 28, 1–17

Elmslie, W.A. (1901) Among the wild Nguni. Oliphant, Anderson & Ferrier, London

Fortes, M. (1940) Children's drawings among the Tallensi. Africa, 13, 239–95

Fortes, M. (1981) Tallensi children's drawings, In Lloyd, B. and Gay, J. (Eds).

Fraser D (1932) The teaching of healthcraft to African women. Longmans, Green & Co., London

Gombrich FH (1962) Art and Illusion, a study in the psychology of pictorial representation. Phaidon Press, London

Goodman N (1972) Languages of art: an approach to the theory of symbols. Hackett, Indianapolis

Gould, C.S. (1994) Clive Bell and Aesthetic Experience and Aesthetic Truth. British Journal of Aesthetics, 34, 124–133

Gregory RL (1968) Visual Illusions. Scientific American, 219: 66–79

Harding JR (1968) Interpretation of White Lady painting of South West Africa, some considerations. South African Archaeological Bulletin 23:31–34

Herskovits, M.J. (1948) Man and his works, Knopf, New York.

Herskovits, M.J. (1959) Art and value. In Redfield, R. Herskovits, M and Ekholm, G.F. (eds.) Aspects of primitive art. Museum of Primitive Art. New York.

Hudson, W. (1962) Cultural problems in pictorial perception. South African Journal of Science, 58, 189–196

Ishizu, T. & Zeki, S. (2013) The brain's specialised systems for aesthetic and perceptual judgement. European Journal of Neuroscience, 37, 1413–1420

Klapper Z.S. & Birch H.G. (1969) Perceptual and action equivalence of photographs in children. Perceptual and Motor Skills, 29, 763–771

Kunnapas TM (1957) Horizontal-vertical illusion and the surrounding field. Acta Psychologica, 13: 35–42

La Gournerie J de (1859) Traité de perspective linéaire contendant les traces pour les tableaux plans and courbes, les bas-reliefs et les decorations theatrales, avec une theorie des effets de perspective. Dalmont et Dunod, Paris (The book is available online under the Guttenberg scheme)

Landor AHS (1883) Alone with the Hairy Ainu, or 3800 miles on a pack saddle in Yezo and a cruise to the Kurile Islands. Murray, London

Livingstone D (1857) Missionary travels and researches in South Africa. John Murray, London

Lloyd AB (1904) Acholi country, Part II. Uganda Notes, 5:18–22

Mountford CP, Edwards R (1962) Aboriginal rock engravings of extinct creatures in South Australia. Man 62: 97–99

Pager H (1974) The magico-religious importance of bees and honey for rock painters and Bushmen in southern Africa. South African Bee Journal, 46: 6–9

Pliny (the Elder) (1952) Natural History, vol. IX (Libri XXXIII-XXXV). Trans. Rackham H, Heinemann, London. also available as: Natural History: a Selection. Trans. Healy JF, Penguin Books, London

Segall MH, Campbell DT, Herskovits MJ (1966) The influence of culture on visual perception. The Bobbs-Merrill Co, Indianapolis

Serpell R, Deręgowski JB (1980) The skill of pictorial perception: an interpretation of cross-cultural evidence. International Journal of Psychology 15: 145–180

Sigel IE, Cocking RR (1977) Cognitive development from childhood to adolescence: a constructivist perspective. Holt, Reinhart and Winston, New York

Sully, J. (1895) Studies of childhood. Longmans Green & Co., London

The New Oxford Dictionary of English (1998) Pearsall J (ed.) Oxford University Press, Oxford

Thomson DC (1902) The Brabizon School of Painters: Corot, Rousseau, Diaz, Millet, Daubigny. Chapman & Hall, London

Thouless RH (1931) Phenomenal regression to the real object. British Journal of Psychology 21: 3399–359

Thouless RH (1932) Individual differences in phenomenal regression. British Journal of Psychology 22: 216–241

Tongue HM (1909) Bushman paintings. Clarendon Press, Oxford
Townley Johnson, R. (1979) (T. Maggs, ed.) Major Rock Paintings of Southern Africa. David Philip, Cape Town.
Leonardo da Vinci *in* Richter JP ed. (1883) The literary works of Leonardo da Vinci compiled from original manuscripts by Jean Paul Richter, Sampson Low, Marston, Searle & Rivington, London
Wade, N. (2016) Art and Illusionists. Springer International Publishing, Switzerland
Waller, B. (1990) Foreword to Csikszentmihalyi and Robinson (1990)
Welford AT (1968) Fundamentals of skill. Methuen, London
Wilson M, Wilson BG (1982) The case of disappearing two-eyed profile: or how little children influence the drawing of little children. Review of Research in Visual Arts Education, 8: 1–8
Witkin, H.A., Dyk, R.B, Faterson, H.F., Goodenough, D.R., Karp, S.A. (1962) Psychological Differentiation. New Your, Wiley
Zeki, S. & Ishizu, T. (2013) The "visual shock" of Francis Bacon: an essay on neuroaesthetics. Frontiers in Human Neuroscience, I, art. 8501

GPSR Compliance

The European Union's (EU) General Product Safety Regulation (GPSR) is a set of rules that requires consumer products to be safe and our obligations to ensure this.

If you have any concerns about our products, you can contact us on ProductSafety@springernature.com

In case Publisher is established outside the EU, the EU authorized representative is:

Springer Nature Customer Service Center GmbH
Europaplatz 3
69115 Heidelberg, Germany

Batch number: 10756378

Printed by Printforce, the Netherlands